Pre-Algebra

Pre-Algebra

Harry H. Jonas
American River College
Sacramento, California

John Wiley & Sons, Inc.
New York London Sydney Toronto

34342

512.9042

Copyright © 1972, by John Wiley & Sons, Inc.

All rights reserved. Published simultaneously in Canada.

No part of this book may be reproduced by any means, nor transmitted, nor translated into a machine language without the written permission of the publisher.

Library of Congress Catalog Card Number: 70-166315
ISBN 0-471-44702-1

Printed in the United States of America.

10 9 8 7 6 5 4 3 2 1

Preface

This book was written with the belief that a fresh approach to a course in Pre-Algebra would be welcomed by many students. There is a genuine need for a text which will actually bridge the gap between arithmetic and algebra. Too many Pre-Algebra texts merely provide familiar arithmetic drills which students find neither stimulating nor rewarding. This approach does little toward improving the probability of student success in a future algebra course. On the other hand, some texts are too formal and demanding, discouraging many students from ever attempting the study of algebra.

This text was based upon the following precepts: A Pre-Algebra text should be easily read by the student by means of an informal, though accurate presentation of the concepts. Students at this level are pragmatic, so that formal arguments go unappreciated. Whenever possible, concepts should be motivated by meaningful examples leading first to an intuitive definition, then to a more formal or symbolic definition. The student should be exposed to abstract symbolism gradually and only when it will simplify the discussion or notation.

Too many authors include topics merely because they are in vogue or to increase the salability of the text. Material should be included only when it can aid in the development of a concept useful to the student at the text level. Ideas presented and not immediately used are too easily forgotten and become meaningless to the student.

The usual arithmetic operations and properties are included here but with a new flavor and organization designed to hold the student's interest. Deductive reasoning is used wherever possible to justify concepts and examples. Students will find this exposure to the scientific method valuable training for later work in algebra and other science courses. There are over 1,900 problems in the text with an adequate number of non-routine problems in each section for at least one homework assignment. Each chapter is summarized and review exercises are provided. The exercises throughout the text duplicate as closely as possible the concepts which must be mastered in algebra, but using the more familiar Hindu–Arabic numerals. Literal symbolism is only used when stating definitions and properties.

Integers are introduced early in the text, not only to develop the skill necessary to work with these numbers in the study of algebra, but as a new and stimulating kind of mathematics for the student. Set concepts and number group properties are introduced gradually and are completely integrated with the development of the real number system.

Each section is intentionally kept short to allow time for discussion of the material and/or frequent short quizzes. The text was written for a junior college class in Pre-Algebra, but would be equally suitable for any mature audience. The author is hopeful that the student will learn some valuable algebraic concepts and in the process develop much needed arithmetic skills. This ability and experience should also provide him with the motivation and confidence necessary to proceed to the study of algebra.

There is more than sufficient material provided here for a one-semester course meeting three days a week. A glossary of terms is provided for a study guide and also for the convenience of the student. The answers to some of the odd-numbered exercises are also included.

Much thanks is given to Mr. Norman E. Andersen for his valuable advice and assistance in class testing the manuscript during the previous year. I am also grateful to Dr. Norman Nystrom, Mrs. Margaret Lial, Mr. Ernest W. Steffen, Mrs. Queen Randall, and other colleagues at American River College, for their suggestions and encouragement. I also wish to express my gratitude to Mrs. Kathleen Seifert and her office staff in the Mathematics Division of American River College for preparing copies of the manuscript for class testing, to the students who endured it in this form, and to Michael Jacobs of Cypress College, Dorothy Tuggle of Denver Community College, and Robert G. Russell of West Valley

Junior College for the constructive criticism they offered for its improvement and clarification. I am indebted to the staff of John Wiley & Sons, in particular Mrs. Phyllis Niklas, Mrs. Elodie Sabankaya, and Mrs. Linda Riffle, all at Palo Alto, California, for their diligent and thorough scrutiny of the text. The myriad of publication details would have overwhelmed me without their competent guidance. Finally, the book could never have been written without the patience and assistance of my wife during this trying period.

Fair Oaks, California
September 1971
                                              HARRY H. JONAS

Contents

1 Whole Numbers

1.1	Simple Grouping Numeral Systems	3
1.2	Hindu–Arabic Numeral System	8
1.3	Addition of Whole Numbers	12
1.4	Addition Algorithm	17
1.5	Subtraction of Whole Numbers	21
1.6	Multiplication of Whole Numbers	27
1.7	Order of Whole Numbers	33
1.8	Summary	38

2 Integers

2.1	Introduction	43
2.2	Addition of Integers	48
2.3	Addition Algorithm	51
2.4	Subtraction of Integers	56
2.5	Multiplication of Integers	61
2.6	Distributive Property	66
2.7	Summary	70

3 Rational Numbers

3.1	Multiplication of Rational Numbers	75
3.2	Fundamental Property of Fractions	84
3.3	Common Factors in Multiplication	91
3.4	Division of Rational Numbers	96
3.5	The Division Rule	102
3.6	Least Common Multiple	106
3.7	Addition and Subtraction of Rational Numbers	109
3.8	Order of Rational Numbers	117
3.9	Compound Fractions	120
3.10	Summary	123

4 Irrational Numbers

4.1	Exponents	129
4.2	Operations with Exponents	137
4.3	Radicals	139
4.4	Operations with Radicals	148
4.5	Rationalizing the Denominator	152
4.6	Evaluation	157
4.7	Summary	162

5 Real Numbers

5.1	Nonpositive Exponents	167
5.2	Decimal Fractions	172
5.3	Operations with Decimals	179
5.4	Scientific Notation	184
5.5	Decimal Approximations for Irrational Numbers	189
5.6	Real Number System	195
5.7	Summary	199

6 Mathematical Sentences

6.1	Algebraic Expressions	205
6.2	Applications of Algebraic Expressions	211
6.3	Equations	215
6.4	Applications of Equations	220
6.5	Inequalities	226
6.6	Summary	231

7 Numeral Systems

7.1	Base Five Numerals	237
7.2	Operations with Base Five Numerals	244
7.3	Numerals Other Than Base Five	251
7.4	Binary Numeral Systems	257
7.5	Summary	263

Table of Squares and Square Roots — 267

Glossary — 269

Answers for Selected Odd-Numbered Exercises — 275

Index — 295

Pre-Algebra

1 Whole Numbers

1.1. Simple Grouping Numeral Systems

Have you wondered about the origins of the number system we use? In this chapter, we shall briefly discuss the history and development of the number system which is such an important part of our daily lives. Numbers are used in every phase of society. It is difficult to think of a business or profession in which numbers do not play a vital role in any given day.

Historians and mathematicians have conjectured that men's early attempts to communicate quantity to each other were in the nature of monosyllabic sounds. Each sound represents a different object or group of objects. Since he had few possessions then, it was only necessary for man to master a small number of sounds. To illustrate:

Group of Objects	Possible Sound
One goat	Ugh
Two goats	Oop
One pig	Yak
Two pigs	Yum

4   Pre-Algebra

This method proved to be too difficult when larger quantities of objects were involved, since it required the memorization of too many different sounds. As civilization became more complex, it became necessary to develop an improved method. Many thousands of years passed before man recognized that one goat, one pig, one wife, and one tree had something in common; that is, "oneness."

Let's reconstruct these ideas in the language of the mathematician. A group of objects having some recognizable property in common is called a **set.** Sets are identified by braces, { }, and also by capital letters. The objects in the set are called **elements** of the set. To illustrate:

(a) $S = \{Mary, Tom, Pete, Sally\}$ — Set of students in my row

(b) $C = \{$Cities in California having a population over 100,000$\}$ — Sacramento, Fresno, and San Diego are some of the elements of this set

(c) $G = \{$Members of the Tom Jones family$\}$ — Mr. and Mrs. Jones, Bob and Sally are the elements of this set

We read (a) as "S is the set of Mary, Tom, Pete, and Sally." Mary is an element of set S. Or alternately, Mary **belongs** to set S, or Mary **is a member** of set S. Sets can be defined by a list of elements as in (a) or by a rule as in (b) or (c). Some sets can only be defined by a rule, some can be defined in both ways, but perhaps more conveniently by one particular method. For example:

(d) $B = \{$Possible sea routes from New York to London$\}$
(e) $C = \{$Cities in the United States having a population of 5,000 or more$\}$

It would be impossible to list the elements of (d), while it is certainly more convenient to state (e) by the rule method.

Now that we have extended our vocabulary, let's go back to prehistoric man. When man noticed that the sets {goat}, {pig}, {wife}, {tree} all had

Whole Numbers 5

the property of oneness in common, he was recognizing that these sets each had one element. The mathematician states this by calling these sets "unit sets," or more important, by saying that these sets each have a **cardinality** of one. Sets having a "twoness" in common, or each having two elements, we say have a cardinality of two. We also say that a set having two elements has a cardinal number of two. Any two sets having the same number of elements are said to have the same cardinality. Specifically, the cardinality of any set is simply the number of elements in that set.

Once man recognized that sets have not only an identity, but a cardinality, he developed new sounds as follows:

Group of Objects	Possible Sound
One goat	Kin ugh
Two goats	Dak ugh
One pig	Kin yak
Two pigs	Dak yak

Note that new and distinct sounds were developed for one, two, goat, and pig. Where twenty distinct sounds were once required to count ten pigs and ten goats, now twelve will suffice: ten sounds for the ten numbers, one sound for goat, and one sound for pig. This was a decided improvement over the previous method.

As the need developed to keep written records of business transactions, tax rolls, and other vital statistics, symbols were invented to replace the sounds. These symbols are called **numerals.** It is important to remember that numerals represent numbers. Thus, the number "two" represents a set having a cardinality of two, and the familiar symbol **2** is the numeral which represents the number two or the idea of "twoness."

One of the earliest known examples of a written numeral system is that of the Egyptian hieroglyphics which has been traced back as early as 3400 BC. Since writing implements were crude at that time, the symbols were by necessity extremely simple.

6   Pre-Algebra

Numerical Value	Symbol	Meaning
1	│	A vertical staff
10	∩	A heel bone
100	୨	A scroll
1,000	⚘	A lotus flower
10,000	⌐	A pointing finger
100,000	⌇	A burbot fish
1,000,000	⚇	An astonished man

If enough of these symbols are grouped together, they will then form the desired number. Thus:

$$23 = \cap\cap|||$$

$$423 = \text{୨୨୨୨}\cap\cap|||$$

$$12{,}351 = \text{⌐⚘⚘୨୨୨}\cap\cap\cap\cap\cap|$$

Here, each symbol has a value regardless of its position in the numeral. Egyptian hieroglyphics and other systems of this type are called **simple grouping systems.** We shall soon see that not all numeral systems are of this type.

A second example of a simple grouping numeral system is the Roman numeral system. Initially only the symbols I, X, C, and M were used, having values of 1, 10, 100, and 1,000, respectively. Later, the intermediate symbols V, L, and D were added, which represent values of 5, 50, and 500, respectively. Examples of Roman numerals are

$$37 = \text{XXXVII}$$
$$562 = \text{DLXII}$$
$$1{,}970 = \text{MDCCCCLXX}$$

Whole Numbers 7

In more recent times, a subtractive principle was used, where 4, 9, 40, 90, 400, 900, and so forth, were more economically written as follows:

$$4 = IV \qquad 90 = XC$$
$$9 = IX \qquad 400 = CD$$
$$40 = XL \qquad 900 = CM$$

Using this principle: 1,970 = MCMLXX.

1.1. Exercises

Write each of the following sets by listing the elements:

1. {Seasons in a year}
2. {Countries which border the United States}
3. {Types of coins now minted by the United States Treasury}
4. {Lakes in the Great Lakes}
5. {Letters in the English alphabet between c and k}
6. {Letters used in the musical scale}

Write each of the following sets by stating an equivalent rule:

7. {New York, Chicago, Los Angeles}
8. {April, June, September, November}
9. {Sunday, Monday, Tuesday, Wednesday, Thursday, Friday, Saturday}
10. {April, May, June}
11. {Washington, Oregon, California}
12. {Pacific, Atlantic, Indian}

State the cardinality of the following sets:

13. {Days of the year}
14. {Hours in a day}
15. {States in the United States}
16. {Presidents of the United States}
17. {□, ○, △, ⬠, ◇}
18. {I, V, X, L, C, D, M}

Write each of the following numbers with Egyptian hieroglyphic numerals:

19. 234
20. 1,041
21. 1,970
22. 20,103

Pre-Algebra

Write each of the following Egyptian hieroglyphic numerals as numbers:

23. ∩∩∩ |||||

24. 𓀀𓀀 999

25. 🐟🐟 ∩∩ |||

26. 👤👤 𓂋𓂋𓂋 𓀀

Write each of the following numbers with Roman numerals:

27. 263   28. 2,364   29. 94   30. 4,044

Write each of the following Roman numerals as numbers:

31. LXII   32. CCXXVI   33. DXC   34. MCMXLI

1.2. Hindu-Arabic Numeral System

Since Roman numerals did not lend themselves well to the mathematical calculations which were emerging during the Renaissance, another system was needed to replace it. In 711 AD, Arabs invading Spain brought with them a numerical system invented in India almost 1,000 years earlier. The Europeans were slow to see the advantages of the so-called Hindu–Arabic numerical system, and it was not generally adopted until the year 1500. Even then, commercial institutions continued to use Roman numerals for almost two more centuries before they accepted the newer system. Roman numerals were now of the past, but it had been the principal numerical system in Europe for over 2,000 years.

    The Hindu–Arabic numeral system uses the set of numerals:

$$\{0,1,2,3,4,5,6,7,8,9\}$$

Each numeral in any given number is called a **digit.** The value of each digit depends upon its position, as well as the value of the numeral. Simply stated, each numeral has **face** and **place** value. In contrast, each digit in a grouping system has face value only. The diagram below

Whole Numbers 9

illustrates the positional method. Starting with **units** on the right and proceeding to the left, each digit in a Hindu–Arabic numeral has a place value ten times greater than the previous digit.

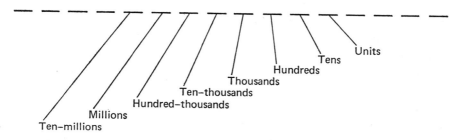

Because each digit possesses positional value, the Hindu–Arabic numeral system is called a **positional** numeral system. The following are examples of Hindu–Arabic numbers:

(a) 272   (b) 4,340   (c) 70,507

In (a), the first digit 2, starting from the left, and the third digit, also 2, both have the same face value. However, the first digit has a place value of 100, while the third digit has a place value of 1. In (b), the first digit 4 has a place value of 1,000, while the third digit 4 has a place value of 10. Note the use of zero in (b). Without it, the number would change identity and become four hundred thirty-four. Again in (c), the 0 preserves the place values of the numerals 7 and 5. The zero places the numerals in their respective positions in the number and for that reason it is sometimes a **placeholder.** Since the zero plays such a vital role in the positional numeral system, it is little wonder that teachers of mathematics become horrified when students refer to zero as "nothing."

We have previously defined number as the cardinality of any given set. Can you conceive of a set whose cardinality is zero? Consider the following sets:

(d) P = {All pole vaulters who can vault over 20 feet}
(e) G = {All people exceeding 10 feet in height}
(f) K = {All people who can individually lift 10 tons of weight}

These sets each have a cardinality of zero and are called **null** sets or **empty** sets. Because of the importance of the null set, it is given the special symbol, $\varnothing$. The empty set of braces, { }, is sometimes used to represent the null set, but the symbol, $\varnothing$, is the more popular of the two and shall be used throughout this text.

Before continuing, it is necessary to introduce some additional mathematical terms that will be useful in what follows. There are two broad varieties of sets: one in which the number of elements can be counted, given enough time, and a second kind where the number of elements cannot be counted. The first type is called a **finite** set and the second an **infinite** set. Cardinality will not be assigned to infinite sets since whole numbers cannot be found which can describe the number of elements in these sets. Consider the following sets:

(g) $B = \{$Possible sea routes from New York to London$\}$
(h) $C = \{$Cities in the United States having a population of 5,000 or more$\}$

Since the number of sea routes in (g) is limitless, it would be impossible to count them. Although there are many cities in the United States over 5,000 in population, one could count them without too much difficulty. Set B is an example of an infinite set, while set C is a finite set. Another example of an infinite set are the numbers 1, 2, 3, 4 and so forth, usually written $\{1,2,3,...\}$, the three dots meaning "and so forth." This set is called the set of **natural numbers,** or **counting numbers,** and it is designated by the capital letter N. Since:

> The cardinality of {goat} is 1
> The cardinality of {goat,cow} is 2
> The cardinality of {goat,cow,horse} is 3, etc.

it can be verified that the elements of N represent the cardinality of any nonempty finite set. The set $\{0,1,2,3,...\}$ includes the natural numbers as well as the number which represents the cardinality of the empty set. This set is called the set of **whole numbers,** and it is designated by the capital letter W. Each number in W represents the cardinality of **any**

Whole Numbers

finite set. Summarizing:

$$N = \{1,2,3,\ldots\} \quad \text{the set of natural or counting numbers}$$
$$W = \{0,1,2,3,\ldots\} \quad \text{the set of whole numbers}$$

1.2. Exercises

Write the Hindu–Arabic numeral for each of the following numbers:

1. Two hundred four
2. Six thousand sixty-six
3. Fifty thousand five hundred five
4. Three hundred twenty-four thousand
5. Four million sixty-three thousand twenty-nine

Write the following numbers with Egyptian hieroglyphic numerals:

6. Twenty-three
7. Ten thousand two hundred twenty

Write each of the following numbers with Roman numerals:

8. Four hundred sixty-nine
9. Nine hundred forty-four

For each of the following Hindu–Arabic numerals, first state the face value, and then the place value of the digits requested:

10. 763; second digit
11. 402,631; second digit
12. 402,631; fifth digit
13. 2,184,137; third digit
14. 67,481,065; second digit

State the cardinalities of each of the given sets:

15. $\{0,1,2\}$
16. $\{0\}$
17. $\emptyset$
18. $\{2,4,6,\ldots,16\}$
19. $\{2,4,6,16\}$
20. $\{2,3,4,\ldots,98\}$

State whether each of the following sets are finite or infinite:

21. $W = \{0,1,2,3,\ldots\}$
22. {Human beings on earth}
23. {Grains of sand which completely fill this classroom}

24. {5,10,15,. . .,100,000,000,000,000}
25. {Possible air routes between Los Angeles and New York}
26. {Two-headed professional basketball players}

Using the three dots wherever possible, describe the following sets by the list method:

27. {All natural numbers between 4 and 176}
28. {Every third natural number from 2 through 35}
29. {Every fifth whole number from 2 through 18}
30. {Every seventh whole number from 4 through 10}
31. {All whole numbers preceding 17}
32. {All natural numbers between 100 and 200}

1.3. Addition of Whole Numbers

The next task is learning to **calculate** using the set of whole numbers just developed. The word "calculate" is defined as "to determine by mathematical processes." The processes which shall be of most concern at this time are **addition, subtraction, multiplication,** and **division,** which are called **binary operations** in the language of mathematics. When two numbers are combined in some way to produce a third number, not necessarily different, the process is called a binary operation. The definition of addition will be developed first by introducing a new set operation. The following sets will serve as examples:

(a) $V = \{a,e,i,o,u\}$

$C = \{c,d,f,g\}$

$B = \{a,c,i\}$

A set operation creating a new set consisting of the elements of V or B, or both, is called the **union** of the two sets, written $V \cup B$:

$$V \cup B = \{a,c,e,i,o,u\}$$

Generally:

> The union of any two sets A and B, written A ∪ B, is the set which consists of all elements which are in either A or B, or both A and B.

Using V, C, and B from example (a),

(b) $B \cup C = \{a,c,d,f,i,g\}$

$V \cup C = \{a,c,d,e,f,g,i,o,u\}$

Observe that sets V and C have no elements in common. When any two sets have no elements in common, they are called **disjoint sets.** Now one final notational device to facilitate our discussion. The cardinality of set A, or the number of elements in set A, can be written $n(A)$.

Thus, from the examples,

$$n(V) = 5 \qquad n(C) = 4 \qquad n(B) = 3$$
$$n(V \cup B) = 6 \qquad n(B \cup C) = 6 \qquad n(V \cup C) = 9$$

Using the plus sign, $+$, to indicate addition, the following relationships are observed (the symbol $\neq$ means "not equal to"):

$$n(V) + n(C) = n(V \cup C) \qquad n(V) + n(B) \neq n(V \cup B)$$
$$n(B) + n(C) \neq n(B \cup C)$$

Note that V and C are disjoint sets while V,B and B,C have elements in common. The previous examples serve to make the following definition plausible:

> The cardinality of the union of two sets A and B is called the **sum** of the cardinalities of A and B, providing sets A and B are disjoint.

The process of finding the sum of two numbers is called **addition.** Each of the numbers in any sum is called an **addend.** Symbolically:

$$n(A) + n(B) = n(A \cup B) \qquad \text{providing A and B are disjoint sets}$$

The definition is sound but easier to use than to state. A second example may make the concept more meaningful:

(c) $D = \{\text{Mary,Tom,Pete}\}$ $\qquad D \cup F = \{\text{Mary,Tom,Joe,Pete, Harry}\}$

$F = \{\text{Joe,Harry}\}$ $\qquad G \cup F = \{\text{Joe,Harry,Pete, Norm}\}$

$G = \{\text{Joe,Pete,Harry,Norm}\}$ $\qquad D \cup G = \{\text{Joe,Pete,Harry, Norm,Mary,Tom}\}$

$H = \{\text{Norm}\}$

Here, $n(D) = 3$, $n(F) = 2$, $n(G) = 4$, $n(D \cup F) = 5$, $n(G \cup F) = 4$, and $n(D \cup G) = 6$, so that:

$$n(D) + n(F) = n(D \cup F) \qquad n(G) + n(F) \neq n(G \cup F)$$
$$3 + 2 = 5 \qquad\qquad 4 + 2 \neq 4$$

$$n(D) + n(G) \neq n(D \cup G)$$
$$3 + 4 \neq 6$$

In example (c), the sum of cardinalities was correct only in the case which involved the disjoint sets D and F. Sets D,G and G,F are not disjoint and so the sum of their cardinalities is not equal to the cardinality of their unions.

Before continuing, two very important properties of addition and union of sets are best introduced at this point using the above example. Note that $F \cup D = D \cup F$ as well as $G \cup F = F \cup G$ and $D \cup G = G \cup D$. The ability to find the union of any two sets in either order without affecting the result is called the **commutative property** of the union operation. Stated symbolically:

For all sets A and B, $A \cup B = B \cup A$.

Now observe in example (c) that $n(D) = 3$ and $n(F) = 2$, so that $3 + 2 = 2 + 3$. Since $n(D) + n(F) = n(F) + n(D)$ holds for the cardinality of any two sets, the operation of addition is also said to be commutative. The commutative property of addition is stated symbolically as

$a + b = b + a$ for any two whole numbers a and b

Referring again to example (c),

$$(D \cup F) \cup G = \{\text{Mary,Tom,Joe,Pete,Harry}\} \cup G$$
$$= \{\text{Mary,Tom,Pete,Harry,Joe,Norm}\}$$

and

$$D \cup (F \cup G) = D \cup \{\text{Joe,Pete,Harry,Norm}\}$$
$$= \{\text{Mary,Tom,Pete,Joe,Harry,Norm}\}$$

so that:

$$(D \cup F) \cup G = D \cup (F \cup G)$$

The ability to group the union of any three sets in two different ways **without** changing the order is referred to as the **associative property** of the union operation. Stated symbolically:

For all sets A, B, and C, $(A \cup B) \cup C = A \cup (B \cup C)$.

In example (c), it is true that

$$[n(D) + n(F)] + n(H) = n(D) + [n(F) + n(H)]$$

It can also be verified that $(7 + 4) + 2 = 7 + (4 + 2)$, and it can be verified generally that the operation of addition is associative. The associative property of addition is stated symbolically as

$(a + b) + c = a + (b + c)$ where a, b, and c are whole numbers

In order to provide the stimulus for another important property, consider the union of the fullowing sets:

$$\{\text{Mary,Tom,Pete}\} \cup \emptyset = \{\text{Mary,Tom,Pete}\}$$

$$\emptyset \cup \{a,c,d,f,g\} = \{a,c,d,f,g\}$$

It can be verified that the union of any set A with the null set, in either order, is equal to the set A. Written symbolically, the statement becomes:

For any set A, $A \cup \emptyset = A$ and $\emptyset \cup A = A$

Since union with the null set, in either order, does not change the identity of any given set, the null set is called the **identity element** for the union operation. In addition,

or
$$n(\{\text{Mary,Pete,Tom}\}) + n(\emptyset) = n(\{\text{Mary,Pete,Tom}\})$$
$$3 \qquad\qquad + \quad 0 \quad = \qquad 3$$

and

so that
$$n(\emptyset) + n(\{a,c,d,f,g\}) = n(\{a,c,d,f,g\})$$
$$0 \quad + \quad 5 \quad = \quad 5$$

Here are two examples of addition of a number with zero, in either order, that does not change the identity of that number. It is also true that addition of 0 with any whole number, in either order, does not

16 Pre-Algebra

change the identity of that number. Written symbolically:

$n + 0 = n$ and $0 + n = n$ where n represents any whole number

Because 0 possesses this property:

0 is called the identity element for addition

1.3. Exercises

Find the union of each of the following sets:

1. $\{2,5,8\} \cup \{3,8,11\}$
2. {Mary,Pete,Alice,Tom} ∪ {Mary,Alice}
3. $\{3,6,9,\ldots,21\} \cup \{15,18,21,\ldots,84\}$
4. {States starting with a C} ∪ {States starting with a W}
5. {Whole numbers 4 through 9} ∪ {Whole numbers 6 through 13}
6. {Boys who have blue eyes} ∪ {Boys who have red hair}
7. $\{a,e,r,s,w\} \cup \emptyset$
8. [{Joe,Harry,Norm} ∪ {Mary,Alice}] ∪ {Joe,Alice}
9. Which of the sets in Problems 1--7 are disjoint sets?

Find the cardinality of each of the following sets:

10. $\{a,e,i,o,u\}$
11. {States starting with a C}
12. $\{3,6,9,\ldots,27\}$
13. $\{3,4,5,6,\ldots,102\}$
14. {Natural numbers greater than 7 and less than 19}
15. $\{3,7,8,17\} \cup \{7,13,17,21\} \cup \{8,13,27\}$
16. $\{4,8,12,\ldots,28\}$

Verify that the union of each of the following sets is commutative, that is, $A \cup B = B \cup A$:

17. $\{a,c,e,g\} \cup \{c,g,h,k\} = \{c,g,h,k\} \cup \{a,c,e,g\}$
18. {Joe,Bob,Pat} ∪ {Bob,Tim,Cathy}
    $=$ {Bob,Tim,Cathy} ∪ {Joe,Bob,Pat}

Verify that the union of each of the following sets is associative, that is, $(A \cup B) \cup C = A \cup (B \cup C)$:

19. $[\{2,7,9\} \cup \{7,11,13\}] \cup \{8,12,13\} = \{2,7,9\} \cup [\{7,11,13\} \cup \{8,12,13\}]$
20. $[\{California, Oregon, Washington\} \cup \{Illinois, Indiana\}] \cup \{Alabama, Georgia, Florida\} = \{California, Oregon, Washington\} \cup [\{Illinois, Indiana\} \cup \{Alabama, Georgia, Florida\}]$

Comment on each of the following statements:

21. If a is an element of A, then a is an element of $A \cup B$.
22. If a is an element of $A \cup B$, then a is an element of A.

Insert the proper word to make each of the following statements true:

23. Since addition can only occur between two numbers, it is an example of a(n) _____ operation.
24. If the sum of the cardinalities of sets C and D is equal to the cardinality of the union of C and D, then C and D are _____ sets.

1.4. Addition Algorithm

In this section a visual or graphical interpretation of addition will be developed to reinforce the definition of addition just presented. The interpretation will be based on the **number line** which shall also be useful in what follows.

On any line which extends indefinitely in both directions, mark a point A. At an arbitrary distance from A, mark a point B. Using the distance from A to B, called the line segment AB, as a unit of measurement, mark off successive units of length equal to AB along the number line; A is called the **graph** of 0 (zero), and B the graph of 1. Since point C is two units to the right of 0, it is called the graph of 2. In this manner, the set of whole numbers can be portrayed. The representation of any set on a number line is called the **graph** of the set. With the aid of the number line, addition can now be easily illustrated:

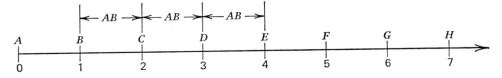

18   Pre-Algebra

To demonstrate that $3 + 2 = 5$, starting from A, mark off a line segment 3 units long, which is AD. From D, mark off a line segment 2 units long, which is DF, as shown.

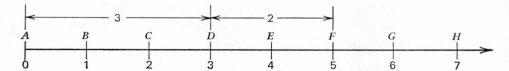

The sum of the numbers is the line segment AF, which is 5 units long.
The diagrams below represent the additional problems shown on the right:

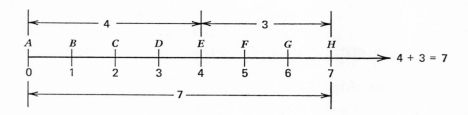

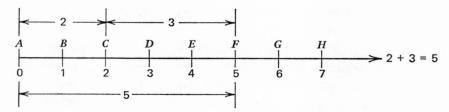

This method illustrates the principle of addition well, but it is not practical to use it with the sum of two numbers such as $263 + 4{,}567$. An easier method is desirable, but before one can be demonstrated, consider the following numbers which have been written in **expanded notation:**

(a) $263 = 200 + 60 + 3$
(b) $4{,}567 = 4{,}000 + 500 + 60 + 7$
(c) $243{,}785 = 200{,}000 + 40{,}000 + 3{,}000 + 700 + 80 + 5$

Using the commutative and associative properties of addition freely,

Whole Numbers 19

the addends are regrouped as shown:

(d) $\quad 47 + 185 = (40 + 7) + (100 + 80 + 5)$
$\qquad\qquad\quad = 100 + (80 + 40) + (7 + 5)$
$\qquad\qquad\quad = 100 + (100 + 20) + (10 + 2)$
$\qquad\qquad\quad = (100 + 100) + (20 + 10) + 2$
$\qquad\qquad\quad = 200 + 30 + 2$
$\qquad\qquad\quad = 232$

Note that the units, tens, and hundreds are regrouped to form the final sum.

Consider a second example:

(e) $\quad 307 + 1,895 = (300 + 7) + (1,000 + 800 + 90 + 5)$
$\qquad\qquad\qquad = 1,000 + (300 + 800) + 90 + (7 + 5)$
$\qquad\qquad\qquad = 1,000 + (1,000 + 100) + 90 + (10 + 2)$
$\qquad\qquad\qquad = (1,000 + 1,000) + 100 + (90 + 10) + 2$
$\qquad\qquad\qquad = 2,000 + (100 + 100) + 2$
$\qquad\qquad\qquad = 2,000 + 200 + 2$
$\qquad\qquad\qquad = 2,202$

Although the method is sound, it is impractical because of the time required. After many centuries, mathematicians have simplified the process. Any mathematical process which has been shortened and simplified is called an **algorithm.**

The example below shows a sequence of steps leading to the algorithm used today:

(f)

(1)	(2)	(3)	(4)	(5)
3 0 7	$\overset{1}{3}$ 0 7	$\overset{1}{3}\overset{1}{0}$ 7	$\overset{1}{3}\overset{1}{0}$ 7	$\overset{1}{3}\overset{1}{0}$ 7
1, 8 9 5	1, 8 9 5	1, 8 9 5	$\overset{1}{1},\,8\,9\,5$	$\overset{1}{1},\,8\,9\,5$
1 2	1 0 2	1 2 0 2	2, 2 0 2	2, 2 0 2
9 0	1 1 0 0	1 0 0 0		
1 1 0 0	1 0 0 0			
1 0 0 0				

20   Pre-Algebra

In (1), the units, tens, hundreds, and thousands are written and added vertically. In (2), the ten resulting from adding the units in (1) is transferred above as indicated. In (3), the one hundred resulting from adding the tens column in (2) is transferred above as indicated. In (4), the one thousand resulting from adding the hundreds column in (3) is transferred above. Finally, the addition algorithm used today is shown in the shortened version (5). Addition in the early grades is generally performed vertically as shown, but it will be advantageous in more advanced courses to also be able to add horizontally.

When adding, whenever possible use number combinations whose sums end in zero. Consider the following sum: $8 + 3 + 2 + 7 + 4 + 2 + 6 = 32$. The commutative and associative properties permit the numbers to be rearranged so that it is far easier to mentally add:

$$(8 + 2) + (3 + 7) + (4 + 6) + 2 = 10 + 10 + 10 + 2$$
$$= 30 + 2$$
$$= 32$$

1.4. Exercises

Graph each of the following sets on a number line:

1. $\{1,3,7\}$
2. $\{0,2,4,6,8,10\}$
3. $\emptyset$
4. $\{1{,}736, 1{,}739, 1{,}742, 1{,}745\}$ (Note: It is not necessary to always start at zero.)

Illustrate each of the following addition facts on a number line by the text method:

5. $3 + 4 = 7$ 6. $5 + 2 = 7$ 7. $2 + 5 = 7$ 8. $3 + 8 = 11$

Write each of the following numbers in expanded notation:

9. 647   10. 4,053   11. 20,807   12. 1,037,460

Find each of the following sums by first writing the number in expanded

Whole Numbers 21

notation and then combining as shown in the text example:

13. 24 + 365   14. 87 + 246   15. 385 + 2,468   16. 3,702 + 2,698

Using the present day algorithm, find the indicated sums. If the problem is written horizontally, perform the addition that way and then check your result by adding again vertically. Repeat vertically written problems by adding horizontally. Use different number combinations each time as a check and also to find the easiest arrangement:

17. 18 + 6 + 4 + 7 + 2   19. 7 + 5 + 9 + 3 + 5 + 4
18. 6   20. 8
    12   14
    24   9
     7   27
    13   6
    —   12
                                       —

Rewrite the following sums vertically, then add using the algorithm:

21. 417 + 2,061 + 23 + 26,704 + 860
22. 19,008 + 6,732 + 17 + 409 + 108,096
23. 3,007 + 430 + 40,007 + 49 + 8,080 + 7
24. 87,003 + 437,534 + 5,806 + 74,738 + 801,108

1.5. Subtraction of Whole Numbers

Questions frequently arise in which it is necessary to find the difference of two numbers in order to arrive at an answer. Consider the following practical examples:

(a) How much more money will you make this year compared with last year's income?
(b) If a board is 17 inches long and 8 inches is cut off, how much of the board remains?
(c) Joe is 6 feet 6 inches tall and Mary is 5 feet 8 inches tall. What is the difference in their heights?

22   Pre-Algebra

The mathematical process of finding the **difference** of two numbers is called **subtraction**. The —, called a **minus** sign, is the symbol used to denote this binary operation. A useful algorithm for subtraction can be developed by first defining subtraction in several ways. A few numerical examples displayed on the number line will aid in arriving at the first definition of this new operation.

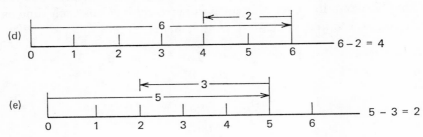

A number line has provided a graphical method for subtracting two whole numbers. A less intuitive method is desirable here and for what follows. Note that in order to find the difference, $6 - 2$, it was necessary to find a number which when added to 2, gave 6. More specifically:

$6 - 2 = c$ if a whole number c can be found to add to 2 to produce 6

Or more generally,

$a - b = n$ if any whole number n can be found so that $a = b + n$

A few examples should clarify this concept:

(f) $7 - 4 = 3$,   since $7 = 4 + 3$
(g) $9 - 5 = 4$,   since $9 = 5 + 4$

In any subtraction problem, the given number is called the **minuend** and the number to be subtracted from the minuend is called the **subtrahend**. In (f), then, the difference 3 can be added to the subtrahend 4 to produce the minuend 7. Generally then:

The difference of any two numbers can be added to the subtrahend to give the minuend.

Whole Numbers 23

Now consider the following examples:

(h) 673 − 431
(i) 475 − 289

The task of finding the whole numbers, representing the difference in each of these examples, is much more difficult than in the previous examples. An algorithm is desirable to simplify the process for these examples as well as for any other subtraction problem.

First, an elementary but important concept must be developed here with the aid of a few examples:

(j) 26 − 12 = 14   By the number line
        = 16 − 2   Since 14 + 2 = 16
        = (26 − 10) − 2   Since 16 + 10 = 26

Now 26 − 12 can be written as 26 − (10 + 2), so that

$$26 − (10 + 2) = (26 − 10) − 2$$

or simply,

$$26 − (10 + 2) = 26 − 10 − 2$$

The parentheses are removed with the provision that the subtractions are performed as originally indicated. That is, from left to right. This shall be commented on again later in this section.

(k) 318 − 172 = 146   By the number line
        = 148 − 2   Since 146 + 2 = 148
        = (218 − 70) − 2   Since 148 + 70 = 218
        = [(318 − 100) − 70] − 2 Since 218 + 100 = 318

If the parentheses and brackets are dropped, then

$$318 − 172 = 318 − 100 − 70 − 2$$

Since 172 = 100 + 70 + 2,

$$318 − (100 + 70 + 2) = 300 − 100 − 70 − 2$$

Summarizing this result:

To subtract a sum from a given number, subtract each of the numbers in the sum from the given number.

24   Pre-Algebra

Now, rewriting (h) and (i) in expanded notation:

(h) $673 - 431 = (600 + 70 + 3) - (400 + 30 + 1)$
$= 600 + 70 + 3 - 400 - 30 - 1$
$= (600 - 400) + (70 - 30) + (3 - 1)$
$= 200 + 40 + 2$
$= 242$

(i) $475 - 289 = (400 + 70 + 5) - (200 + 80 + 9)$
$= 400 + 70 + 5 - 200 - 80 - 9$
$= (400 - 200) + (70 - 80) + (5 - 9)$
$= (400 - 200) + (60 - 80) + (15 - 9)$
$= (300 - 200) + (160 - 80) + (15 - 9)$
$= 100 + 80 + 6$
$= 186$

Observe in example (i) that in order to subtract 9 from 5, it was necessary to "borrow" 10 from 70. In order to subtract 80 from 60, 100 was borrowed from 400 making it 300. Subtraction, using expanded notation, produces the desired difference, but the method is far too lengthy. A shorter method can now be demonstrated, again using example (h).

$$
\text{(h)} \quad \begin{array}{r} 673 \\ -431 \\ \hline 242 \end{array} = \begin{array}{r} 600 \\ -400 \\ \hline 200 \end{array} + \begin{array}{r} 70 \\ -30 \\ \hline 40 \end{array} + \begin{array}{r} 3 \\ -1 \\ \hline 2 \end{array}
$$

Note that the hundred, ten, and unit digits of each number are subtracted from each other. A second example using (i) is reviewed:

$$
\text{(i)} \quad \begin{array}{r} 475 \\ -289 \\ \hline \end{array} = \begin{array}{r} 400 \\ -200 \\ \hline (\ \ ) \end{array} + \begin{array}{r} 70 \\ -80 \\ \hline (\ \ ) \end{array} + \begin{array}{r} 5 \\ -9 \\ \hline (\ \ ) \end{array}
$$

$$
\begin{array}{r} 4\,{}^{6}\!\!\!\not7\,{}^{1}5 \\ -2\ 8\ 9 \\ \hline 6 \end{array} = \begin{array}{r} 400 \\ -200 \\ \hline (\ \ ) \end{array} + \begin{array}{r} 60 \\ -80 \\ \hline (\ \ ) \end{array} + \begin{array}{r} 15 \\ -\ 9 \\ \hline 6 \end{array}
$$

$$
\begin{array}{r} {}^{3}\!\!\!\not4\,{}^{16}\!\!\!\not7\,{}^{1}5 \\ -2\ 8\ 9 \\ \hline 1\ 8\ 6 \end{array} = \begin{array}{r} 300 \\ -200 \\ \hline 100 \end{array} + \begin{array}{r} 160 \\ -\ 80 \\ \hline 80 \end{array} + \begin{array}{r} 15 \\ -\ 9 \\ \hline 6 \end{array}
$$

Note that the same "borrowing" was done here, but much more compactly than in the expanded notation form. This last method is the subtraction algorithm now in use after many centuries of improvements.

Now consider an example involving several subtractions:

(1) $9 - 6 - 2$

First solution:
$$9 - 6 - 2 = (9 - 6) - 2$$
$$= 3 - 2$$
$$= 1$$

Second solution:
$$9 - 6 - 2 = 9 - (6 - 2)$$
$$= 9 - 4$$
$$= 5$$

Obviously, two solutions are one more than can be tolerated. However, which of the two is correct? To avoid this confusion, the first solution is generally accepted as the correct method, and the rule is adopted that:

> When one or more subtractions are included in any given statement, perform these from left to right, unless grouping symbols such as parentheses indicate otherwise.

When more than one operation is included in any one given statement, it is referred to as a **mixed operation.** Statements of this type also require special attention, as illustrated with the following example:

(m) $7 - 4 + 2$

First solution:
$$7 - 4 + 2 = (7 - 4) + 2$$
$$= 3 + 2$$
$$= 5$$

Second solution:
$$7 - 4 + 2 = 7 - (4 + 2)$$
$$= 7 - 6$$
$$= 1$$

Here again, which is the correct solution? Mathematicians have agreed to resolve this dilemma by formulating a statement around the first solution consistent with the rule developed for multiple subtractions.

That is:

> Mixed operations, involving additions and/or subtractions, are performed from left to right, unless otherwise indicated by grouping symbols such as parentheses.

Finally, not all differences involving whole numbers can be found. Differences such as: $2 - 6$, $9 - 11$, and $1 - 3$ have no solution in the set of whole numbers. The mathematical term describing this concept is called **closure.** It is said that:

> The set of whole numbers is not **closed** with respect to subtraction.

Or:

> The set of whole numbers does not have **closure** under subtraction.

Whenever any two numbers in a given set are combined by a given operation, with the result always another member of the given set, that set is said to be closed with respect to the given operation.

Since the sum of any two natural numbers is again a natural number, the set of natural numbers, N, is closed with respect to addition. The set of whole numbers is also closed with respect to addition for the same reason. The finite subset of N, $\{1,2,3,4,5\}$, is not closed with respect to addition, because there exists at least one sum, $4 + 3$, which is not an element of this set. In fact, there is no finite subset of the set of natural numbers which is closed with respect to addition or subtraction.

The set of whole numbers is not closed with respect to subtraction, but it is highly desirable that a set exists which is. The student shall see the development of such a set of numbers in the next chapter.

Some final observations:

> Since $6 - (4 - 2) \neq (6 - 4) - 2$, subtraction is not an associative operation.
> Since $6 - 4 \neq 4 - 6$, subtraction is not a commutative operation.

1.5. Exercises

Find each of the following differences by referring to the number line. If no whole number can be found, then write "none exists":

1. $17 - 9$
2. $38 - 24$
3. $6 - 8$
4. $12 - 2 - 6$

Whole Numbers

Using expanded notation, find the difference of each of the following statements:

5. $47 - 26$
6. $246 - 29$
7. $3,863 - 479$
8. $5,248 - 674$

Subtract each of the following as indicated, using the algorithm developed in Section 1.5, p. 24:

9. $248 - 169$
10. $4,268 - 2,629$
11. $67,403 - 42,817$
12. $128,093 - 88,706$

Simplify each of the following expressions which include mixed operations. Express your answer as a single whole number. If no whole number can be found then write "none exists":

13. $17 - 3 - 8$
14. $12 + 6 - 4$
15. $28 - 9 + 4 - 12$
16. $17 - 12 + 7 - 9$
17. $23 - 17 - 8$
18. $183 - 49 - 83 + 208 - 243$
19. $(19 - 12) - 5$
20. $19 - (12 - 5)$
21. Is there a finite subset of the set of whole numbers which is closed with respect to addition? With respect to subtraction?
22. Is there a finite subset of the set of natural numbers which is closed with respect to addition? With respect to subtraction?
23. Is there an infinite subset of the set of natural numbers which is closed with respect to addition? With respect to subtraction?

1.6. Multiplication of Whole Numbers

Repeated additions of the same number occur so frequently when performing calculations, that their sums are more easily memorized than added each time they are encountered. For example:

$2 + 2 + 2 + 2$ can be written $4 \cdot 2$ and then memorized as 8
$6 + 6 + 6 + 6 + 6$ can be written $5 \cdot 6$ and then memorized as 30

The mathematical process of repeated addition is called **multiplication,** and the result is called the **product.** The dot and parentheses are used in different ways to indicate multiplication. For example:

$$4 + 4 + 4 = 3 \cdot 4 = (3)(4) = 3(4) = (3)4$$

28   Pre-Algebra

The letter x is a symbol frequently used in Algebra to represent an unknown quantity, and therefore it is seldom used to indicate multiplication. One exception to this rule is presented in Chapter 5. Generally then, the sum of a b's is written:

$$\underbrace{b + b + b + \cdots + b}_{a \text{ addends}} = a \cdot b$$

The equalities $4 \cdot 2 = 2 \cdot 4$ and $8 \cdot 7 = 7 \cdot 8$ verify that multiplication can be performed in either order and therefore these products are commutative. The equalities $4 \cdot (2 \cdot 3) = (4 \cdot 2) \cdot 3$ and $(7 \cdot 8) \cdot 5 = 7 \cdot (8 \cdot 5)$ verify that the product of three numbers remains the same regardless of the way in which they are grouped, and therefore, multiplication is associative. These two properties cannot be verified in general, but are assumed to be true and are symbolized as follows:

$$a \cdot b = b \cdot a$$
$$a \cdot (b \cdot c) = (a \cdot b) \cdot c$$

where a, b, and c are whole numbers

That is, multiplication of whole numbers is commutative and associative.

Multiplication of single-digit numbers are easily memorized, but an algorithm is necessary to find products involving numbers containing more than one digit. Before developing this algorithm, first consider the following multiplications performed with the aid of expanded notation:

(a) $4 \cdot 26 = 26 + 26 + 26 + 26$
$= (20 + 6) + (20 + 6) + (20 + 6) + (20 + 6)$
$= (20 + 20 + 20 + 20) + (6 + 6 + 6 + 6)$ Using commutative and associative properties

$= 4 \cdot 20 + 4 \cdot 6$
$= 80 + 24$
$= 80 + (20 + 4)$
$= (80 + 20) + 4$ Using the associative property

$= 100 + 4$
$= 104$

(b) $73 \cdot 28 = \underbrace{28 + 28 + 28 + \cdots + 28}_{73}$

$\underbrace{(20 + 8) + (20 + 8) + \cdots + (20 + 8)}_{70} + \underbrace{(20 + 8) + \cdots + (20 + 8)}_{3}$

$= 70 \cdot 20 + 70 \cdot 8 + 3 \cdot 20 + 3 \cdot 8$
$= (7 \cdot 10) \cdot (2 \cdot 10) + (7 \cdot 10) \cdot 8 + 60 + 24$
$= 14 \cdot 100 + 56 \cdot 10 + 60 + 24$
$= 1{,}400 + 560 + 60 + 24$
$= (1{,}000 + 400) + (500 + 60) + 60 + (20 + 4)$
$= 1{,}000 + (400 + 500) + (60 + 60 + 20) + 4$
$= 1{,}000 + 900 + (100 + 40) + 4$
$= 1{,}000 + (900 + 100) + 40 + 4$
$= 1{,}000 + 1{,}000 + 40 + 4$
$= 2{,}000 + 40 + 4$
$= 2{,}044$

These examples are designed to demonstrate that multiplication can be performed by regrouping into tens, hundreds, thousands, etc. Obviously, the method of expanded notation is too long, and a shorter algorithm is desired.

Now consider example (a) again:

$$
\begin{array}{cccccc}
 & (1) & & (2) & (3) & (4) \\
26 & 20 & 6 & 26 & \overset{2}{26} & \overset{2}{26} \\
\underline{4} = & \underline{4} + & \underline{4} = & \underline{4} & \underline{4} & \underline{4} \\
 & 80 & 24 & 24 & 4 & 104 \\
 & & & 80 & 100 &
\end{array}
$$

In (1), 26 is separated into $20 + 6$ and each is multiplied by 4. In (2), the products are written vertically. The 2 tens are transferred above in (3); then the 2 tens and 8 tens are added to give 10 tens or one hundred in (3). In (4), the products are added to give the final result.

Now consider example (b) again:

```
        (1)           (2)         (3)          (4)
                                   ²                ²
 73    70 + 3       70 + 3       73      73       73
 28  =   20     +     8    =     28  ⇐   28   =   28
 ──    ─────        ─────        ──      ──       ──
         60           24         24⎫               584
        1400          560        560⎭----584
                                 60⎫              146
                                 1400⎭----1460   ─────
                                                  2,044
```

In (1), 20 is multiplied by 70 + 3 and 8 is multiplied by 70 + 3. The results from (1) are added vertically in (2). The 2 tens from the product of 8 and 3 in (2) are transferred above as indicated in (3). The product of 20 · 73 is added and also written as indicated in (3). The 0 in the unit column is omitted as indicated and the final result is written in (4).

Now, a third and last example:

(c)

```
        (1)           (2)          (3)          (4)        (5)
                                    ²            ⁴
 48    40 + 8      40 + 8          48           48          48
 63  =   60    +    3       =      63    ⇐      63    =     63
 ──    ─────       ─────           ──           ──          ──
         60          3             24⎫                      144
        1400        ───            120⎭----144   144        288
                                   480          480⎫       ─────
                                   2400         2400⎭----2880
                                                            3,024
```

In (1), 48 is multiplied by 60, and 48 is multiplied by 3. The results from (1) are written vertically in (2). The 2 tens from the product of 8 and 3 are transferred above as indicated in (3). The 4, which is the hundreds digit from the product of 60 and 8, is transferred above in (4), as indicated. Again, in (5), the final form is written in which the 0 is omitted.

The final multiplication algorithm required many refinements before it arrived in the twentieth century as it is now known.

Before completing this section, several important properties of multiplication are presented. Considering that $7 \cdot 1 = 7$ and $1 \cdot 7 = 7$, $3 \cdot 1 = 3$

Whole Numbers 31

and $1 \cdot 3 = 3$, it should be clear that the identity of any given natural number is unchanged when it is multiplied by 1 in either order. Therefore:

1 is the identity element of multiplication

Next, it is clear that the sum of repeated addition of 0 is 0, so that:

$$0 + 0 + 0 + 0 = 4 \cdot 0 = 0$$

and by the commutative property,

$$4 \cdot 0 = 0 \cdot 4 = 0$$

This seemingly simple but very important fact is sometimes called the **zero factor property**. This property can be stated as:

> The product of zero and any other whole number is zero. Or symbolically:
>
> $a \cdot 0 = 0 \cdot a = 0$ for any whole number a

Finally, calculations involving mixed operations require clarification. Consider the following examples:

(d) $6 \cdot 4 + 3$

    First solution: $(6 \cdot 4) + 3 = 24 + 3$
$$= 27$$

    Second solution: $6 \cdot (4 + 3) = 6 \cdot 7$
$$= 42$$

(e) $26 - 4 \cdot 3 + 2$

    First solution: $(26 - 4) \cdot 3 + 2 = 22 \cdot 3 + 2$
$$= 66 + 2$$
$$= 68$$

    Second solution: $26 - (4 \cdot 3) + 2 = 26 - 12 + 2$
$$= 14 + 2$$
$$= 16$$

    Third solution: $26 - 4 \cdot (3 + 2) = 26 - 4 \cdot 5$
$$= 26 - 20$$
$$= 6$$

32 Pre-Algebra

Note that three different solutions were obtained in (e) and two solutions were found in (d), depending upon the order in which the operations were performed. In order to eliminate this confusion, mathematicians have agreed to standardize upon the following convention:

> When multiplication, along with addition and/or subtraction, occurs in any given expression, all multiplications are performed first, providing that grouping symbols such as parentheses do not indicate otherwise. The remaining additions and subtractions are then performed, proceeding from left to right, as indicated in Section 1.5, p. 26.

In simplifying example (d), the first solution is the one which is correct. The second solution in example (e) yields the correct answer.

1.6. Exercises

Using the present day multiplication algorithm, find each of the following products:

1. $7 \cdot 27$
2. $38 \cdot 69$
3. $42 \cdot 308$
4. $523 \cdot 64$
5. $372 \cdot 438$
6. $6{,}381 \cdot 29$
7. $236 \cdot 47 \cdot 0$
8. $46 \cdot 38 \cdot 203$
9. $42 \cdot 0 \cdot 138$
10. $1{,}002 \cdot 4{,}003$

Perform the mixed calculations as indicated in each of the following:

11. $7 \cdot 8 + 5$
12. $7 + 3 \cdot 9$
13. $9 \cdot 6 - 8$
14. $3 \cdot 8 + 7 \cdot 6$
15. $23 - 3 \cdot 6$
16. $4 + 3 \cdot 5 - 8$
17. $6 \cdot 6 - 4 + 9$
18. $3 \cdot 7 + 6 + 6 \cdot 7$
19. $13 - 4 \cdot 3 + 15$
20. $43 - 16 - 3 \cdot 8$

21. $482 - 13 \cdot 17 + 4 \cdot 19$
22. $28 \cdot 37 - 13 \cdot 48 - 3 \cdot 93$
23. $5(7 + 4) - 7(8 - 2)$
24. $(9 + 7)(3 + 4) - (3 + 8) \cdot 9$
25. $(9 - 3) \cdot 7 - 3(4 + 7)$
26. $66 - 7(6 + 2) + 3(4 + 9) - 5 \cdot 3$
27. $7 \cdot 6 - 5 \cdot 4 - 3 \cdot 2$
28. $23 - (8 - 5)(7 - 3)$
29. $(17 - 12)(8 - 3)(14 - 6)(12 - 8 - 4)$
30. $17 \cdot (8 - 5) - (12 - 4) \cdot 6 + 5 \cdot (6 - 2)$

1.7. Order of Whole Numbers

Most people today are concerned at one time or another with comparing numerical quantities of almost everything conceivable. Consider the following examples of statements frequently heard:

(a) Sally's weight is greater than Mary's.
(b) Jim's car cost less than Tom's.
(c) Pete high jumped less than 6 feet.
(d) The population of Los Angeles is greater than that of Sacramento.
(e) Apex Company's gross profit is the same this year as last year.
(f) The maximum speed of car A is the same as that of car B.

In each of the above statements, two quantities are compared. In statements (a), (b), (c), and (d), the quantities are not the same; they are unequal. The quantities in statements (e) and (f) are the same; they are equal. Statements in which the quantities compared are equal are called **equalities**. Those statements where the quantities compared are unequal are called **inequalities**. The symbol used to indicate equality is the familiar equals sign, $=$. Two symbols are used to show inequality: $>$ to indicate "greater than," and $<$ for "less than" statements. The following examples will clarify the use of these new symbols:

(g) $6 > 2$ is read as "six is greater than two"
(h) $8 < 11$ is read as "eight is less than eleven"
(i) $0 < 4$ is read as "zero is less than four"

34 Pre-Algebra

(j) $n(\{Joe, Mary, Steve\}) > n(\{sheep, cow\})$ is read as "the cardinality of the first set is greater than the cardinality of the second set" (specifically, $3 > 2$)

The examples provided here can easily be verified by intuition, or experience. Later in the text this may not be possible and it will be advantageous to have a more systematic approach to assist with these kinds of problems. We offer two methods, one using the number line, the other involving addition.

The number line below shows the graphs of the numbers 4 and 7:

Since a set having a cardinality of 7 has more elements than one having a cardinality of 4, it holds that 7 is greater than 4, or $7 > 4$. Also observe that 7 is to the right of 4 on the number line. It can easily be verified that 4 is to the right of 0 so that $4 > 0$. These facts can be generalized so that a less intuitive definition of inequality can be stated.

Any whole number is greater than a second if the graph of the first number is to the right of the graph of the second on the number line. Stated symbolically:

$b > a$ if b is to the right of a on the number line

Note in the example above that 7 is three units to the right of 4, or, $4 + 3 = 7$. In other words, a natural number was added to 4 to produce 7. Again, since 4 is four units to the right of 0, 4 can be added to 0 to give 4, or $0 + 4 = 4$. In both cases, a natural number was added to the smaller to produce the larger number. This can be stated as an alternate definition of inequality:

$b > a$ if $a + n = b$ where n is a natural number

Restating this definition without the seemingly complicated symbolism:

> For any two whole numbers, one is greater than a second, if a natural number can be found to add to the smaller number to produce the larger.

Whole Numbers   35

Rather than formulate a new definition for the "less than" inequality, it is simpler to define "less than" in terms of "greater than." For example:

$$\text{If} \quad 7 > 4, \quad \text{then} \quad 4 < 7.$$
$$\text{If} \quad 4 > 0, \quad \text{then} \quad 0 < 4.$$

Generalizing, it can be said that:

$$\text{For any two whole numbers, if} \quad a > b, \quad \text{then} \quad b < a.$$

Frequently, statements are heard in which the quantities are compared in terms which are equivalent to the phrases "less than or equal to" or "greater than or equal to." The following examples illustrate these cases:

(k) Joe plans this year to make at least as much money as he did last year.
(l) Joe's income this year will be greater than or equal to last year's income.

Continuing the examples:

(m) Irene weighs no more than Sandy.
(n) Irene's weight is less than or equal to Sandy's.

Statements (k) and (m) have been restated as statements (l) and (n), respectively. Note that the latter statements sound odd using the mathematical terminology because few people speak with such precision.

The phrase "less than or equal to" is symbolized as $\leq$, while the phrase "greater than or equal to" is symbolized as $\geq$.

Consider the following statements which have been written as inequalities on the right of the page. In each case, b represents a whole number:

(o) Any whole number less than 16 $b < 16$ or $16 > b$
(p) Any whole number greater than 7 $b > 7$ or $7 < b$
(q) Any whole number greater than 7 and less than 16 $b > 7$ and $b < 16$
      or $7 < b$ and $b < 16$
      or $7 < b < 16$
(r) Any whole number between 7 and 16 $7 < b < 16$
(s) Any whole number greater than or equal to 4 $b \geq 4$ or $4 \leq b$

(t) Any whole number less than or equal
to 12 $\qquad b \leq 12 \quad \text{or} \quad 12 \geq b$
(u) Any whole number from 4 through 12 $\qquad b \geq 4 \quad \text{and} \quad b \leq 12$
$\qquad\qquad \text{or} \quad 4 \leq b \quad \text{and} \quad b \leq 12$
$\qquad\qquad \text{or} \quad 4 \leq b \leq 12$

Note that the inequalities in (o) and (p) have been combined to produce the double inequalities (q) and (r). Inequalities (s) and (t) have been rewritten as (u). Double inequalities are only used when a set of numbers is **between** two fixed numbers.

A double inequality can be written in any of the eight following ways, where a, b, and c represent whole numbers:

$a < b < c$ is read as		a is less than b and b is less than c
	or	b is greater than a and less than c
$a \leq b < c$ is read as		a is less than or equal to b and b is less than c
	or	b is greater than or equal to a and less than c
$a < b \leq c$ is read as		a is less than b and b is less than or equal to c
	or	b is greater than a and less than or equal to c
$a \leq b \leq c$ is read as		a is less than or equal to b and b is less than or equal to c
	or	b is greater than or equal to a and less than or equal to c

The inequalities (o), (p), (s), and (t) are sometimes referred to as **single**, or **simple** inequalities to distinguish them from the double inequalities. Inequalities of either type are frequently graphed to visualize them more easily.

In Section 1.3, addition was introduced which is an operation upon two numbers. In contrast, this section has included a discussion of equality and inequality, which are **relations** between two numbers.

Whole Numbers 37

1.7. Exercises

Insert the proper symbol ($>$, $<$, $=$) in each of the following blanks in order to state the proper relationship between the two numbers:

1. 3 ____ 1
2. 17 ____ 27
3. $n(\{a,c,f,g\})$ ____ $n(\{e,h,j\})$
4. $4 + 6$ ____ 9
5. 11 ____ $8 + 3$
6. 29 ____ 92
7. $n(\{\text{whole numbers less than } 9\})$ ____ $n(\{\text{natural numbers less than } 9\})$
8. $8 + 3 + 7 + 5$ ____ $9 + 6 + 4 + 2$
9. $48 + 237 + 5{,}038 + 803$ ____ $4{,}863 + 762 + 403 + 84$
10. $n(\varnothing)$ ____ $n(\{0\})$

State whether each of the following statements are true or false:

11. $7 < 5$
12. $0 \leq 3$
13. $8 + 3 > 7 + 4$
14. $9 + 8 \leq 8 + 9$

Write each of the following statements as an inequality:

15. Any natural number greater than 8 and less than 13
16. Any whole number between 9 and 27
17. Any natural number less than 9
18. Any natural number greater than 13
19. Any whole number from 17 through 83

Complete the following statements:

20. $17 > 8$, because ____ can be added to 8 to produce 17
21. The cardinality of $\varnothing$ is ____ (than) the cardinality of any other set.

Graph each of the following sets on a number line:

22. {Whole numbers less than 7}
23. {Natural numbers from 12 through 17}
24. {Whole numbers between 43 and 51}

38   Pre-Algebra

Complete the following statement:

25. $2 > 5$ is false because _____

1.8. Summary

The material presented in this chapter has been planned to give the reader some insight into the origins and development of the numeral system used today. The brief excursion into the several simple grouping numeral systems was made with the hope that a much better understanding and appreciation of the Hindu–Arabic numeral system would ensue. Some mathematical devices and notation were introduced to better explain the present topics as well as to prepare the reader for what follows in successive chapters.

The introduction of the whole numbers was also woven into the chapter along with the binary operations of addition, subtraction, and multiplication of whole numbers. The commutative and associative properties, and the identity element for addition were presented, all of which are important basic properties, vital for a better comprehension of the mathematical calculations which will follow.

Finally, a distinction was made between equality and inequality, along with definitions of the standard symbols used for each of these relations.

1.8. Review Exercises

Write each of the following as Hindu–Arabic numerals:

1.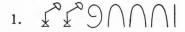

2. CDXXIV
3. MMLXIX
4. $n(\{1,4,7,\ldots,28\})$
5. Two hundred thousand two hundred two

List the elements of each of the following sets; use dots wherever convenient:

6. {Every fourth whole number starting with 0 and less than 19}
7. {All natural numbers between 8 and 93}

Whole Numbers 39

8. {Every third whole number starting with 23, each larger than the previous number}

Write the union of each of the following sets:

9. {a,c,d,g} ∪ {g,r,t,z}
10. {0,4,8,12,...} ∪ {2,6,10,14,...}
11. {17,18,19,...,83} ∪ {26,27,31,47}
12. {Students at American River College registered for more than 12 units} ∪ ∅

Write each of the following numbers in expanded notation:

13. 437
14. 43,078

Perform the indicated multiplication by algorithm:

15. $74 \cdot 23$
16. $4,835 \cdot 267$
17. $284 \cdot 0 \cdot 87$
18. $37 \cdot 58 \cdot 94$

Perform each of the following mixed calculations:

19. $7 \cdot 5 + 4 \cdot 9$
20. $15 - 8 + 3 - 10$
21. $27 - 6 \cdot 4 + 9 \cdot 7 - 59$
22. $183 - 7 \cdot (4 + 8) - 5 \cdot (7 - 3)$

Write each of the following statements as an inequality:

23. Any natural number between 5 and 73
24. Any whole number starting with 17 up to but not including 97
25. Any whole number larger than 43
26. Any natural number less than or equal to 87

Supply the missing word or words to make the following statements true:

27. The statement $3 < 4$ is called a(n) _____
28. The set {1,2,3,...} is called the set of _____ numbers
29. The number of elements in a set is called the _____ of the set
30. The numeral system in use in the U.S. today is called the _____ _____ numeral system
31. Addition is best described as a(n) _____ _____

Pre-Algebra

32. If two sets have no elements in common, they are called _____ sets
33. The symbols used in the number 376 are called _____
34. The number 475 consists of three _____
35. The Hindu–Arabic numeral system is an example of a(n) _____ system whereas the Roman numeral system typifies a(n) _____ _____ numeral system

2 Integers

2.1. Introduction

There are many pairs of whole numbers which cannot be subtracted to yield another whole number. The differences $2 - 6$ and $4 - 5$ cannot be found in the set of whole numbers and therefore it is said that this set is not closed with respect to subtraction. The following discussion provides a justification for the creation of a set of numbers which is closed with respect to subtraction.

A distinguishing characteristic of the graph of the set of whole numbers is that no number exists to the left of the reference point, zero. That is, no whole number exists which is less than zero. In this chapter, we justify the creation of numbers which do exist to the left of the number zero on the number line. Consider the following sets of numbers which are frequently used:

(a) Centigrade temperatures
(b) Altitudes
(c) Financial profit and losses

44   Pre-Algebra

The sets of numbers in these examples all differ from the set of whole numbers, in that, although they all have a reference point, they do not start there. Centigrade temperatures, in (a), can be above or below 0°. Altitudes, in (b), exist above or below zero elevation, or sea level. Profits, in (c), are above the break-even point, zero, whereas losses are below zero. These examples demonstrate the practicality of having sets of numbers which extend in both directions from the reference point, zero. Using the whole numbers as a nucleus, a similar set of numbers can be created. These new numbers are called the set of **integers** and their usefulness will soon become apparent.

When developing the number line in Chapter 1, the right end of the line segment AB was defined as the graph of the number 1, when point A is at 0. By sliding AB successively to the right, the graph of the set of whole numbers was created. Now slide the line segment AB in the **opposite** direction, to the left of zero, so that the right end of AB coincides with A, and call the left end point B' (read B prime). The point B' is the graph of the opposite of one, called **negative** one, and written -1. By continuing to slide the line segment AB successively to the left and marking points C', D', etc., the set of negative numbers can be generated as shown on the number line below:

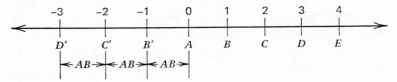

The negative, or opposite of zero, is zero. That is, $-0 = 0$. Numbers to the right of 0 are not negative, and are therefore called **positive** numbers to stress that fact. In the new extended set then, natural numbers become **positive integers.** The following chapters shall reveal still more positive numbers that are not integers. Positive numbers are sometimes preceded by a "+" sign for emphasis, but the value of the number is unchanged if the sign is omitted. For example, $+2 = 2$, $7 = +7$, $+0 = 0$. The + sign in the number $+2$ is read **positive** 2, whereas the same sign in the sum $7 + 2$ is read 7 **plus** 2. The first usage of + indicates a direction on the number line, the second an operation. Summarizing this information in set notation:

  Positive integers = {1,2,3,...} = {+1,+2,+3,...}
  Negative integers = {...,−3,−2,−1}

The union of the sets of negative integers, positive integers, and zero, is called the set of integers, and is designated by the letter *J*. Stated symbolically:

  J = {...,−3,−2,−1} ∪ {0} ∪ {1,2,3,...}

or

  J = {...,−3,−2,−1,0,1,2,3,...}

Note that all three sets, {positive integers}, {negative integers}, and {0}, are disjoint, and that 0 is neither a positive nor a negative integer. The union {0} ∪ {1,2,3,...}, which is {0,1,2,3,...}, is called the set of **nonnegative** integers. The union {...,−3,−2,−1} ∪ {0} is equivalent to {...,−3,−2,−1,0}, and is called the set of **nonpositive** integers.

It should be clear that every natural number is a whole number, and every whole number is an integer. This fact can be stated mathematically that the set of natural numbers is a **subset** of the set of whole numbers, and the set of whole numbers is a subset of the set of integers. In general:

> If each element of any set A is also an element of a set B, then set A is called a subset of set B and this is symbolized as $A \subseteq B$.

Therefore, $W \subseteq J$, $N \subseteq J$, and $N \subseteq W$.

The following examples illustrate this new concept:

(d) {2,7} ⊆ {1,2,3,5,7,9}
(e) {3,5} ⊆ {3,5}
(f) ∅ ⊆ {1,2,3,4,5}
(g) Given the set {1,2},
   ∅ ⊆ {1,2}   {1} ⊆ {1,2}   {2} ⊆ {1,2}   {1,2} ⊆ {1,2}

Although both sets are identical in (e), it is nevertheless true that every element of the first set is an element of the second set. In example (f), there are no elements in the null set, so it is trivially true that every element of ∅ is an element of {1,2,3,4,5}. In fact, the null set is a subset of every set for the very same reason. In (g), the four sets listed are all of the subsets of {1,2}. Note that subset is a relation (between sets) and not an operation.

In order to discover how integers are added, it will be necessary to introduce two concepts—**distance** and **directed distance**—and to learn the difference between them. This can best be accomplished by using the number line and a few examples. The two numbers, -2 and 2, represent different quantities, yet each are the same distance from 0 on the number line. The same is true for -7 and 7, -13 and 13, and in fact it is true for any two numbers which only differ by having opposite signs. Mathematicians say that two numbers have the same **absolute value** when they only differ in sign. The symbol for absolute value is a vertical line on each side of the number. For example:

and
$\quad$ the absolute value of -3 is $|-3|$
$\quad$ the absolute value of -7 is $|-7|$

Since -2 and 2 each have the same absolute value, $|-2| = 2$, whereas $-2 \neq 2$. Both -2 and 2 are the same distance from 0, **without regard to direction,** and since they also have the same absolute value, it seems natural to assert that distance from 0 and absolute value are essentially the same. Repeating:

> The absolute value of a number represents the distance of that number from zero, without regard to direction.

Since -2 represents a distance of 2 units to the left of 0 on the number line, it is called a **directed distance.** Likewise, 2 or $+2$ represents a directed distance of 2 units to the right of 0 on the number line. Therefore, $|-2|$ merely represents a distance on the number line, while -2 represents a directed distance.

In view of the above definition, $|-2|$ and $|+2|$ both represent a distance of 2 units from 0 so that $|-2| = 2$ and $|+2| = 2$. Simply stated, the absolute value of any nonzero number is a positive number. Since 0 is zero distance from 0, $|0| = 0$, or the absolute value of zero is zero.

In concluding this section, observe that -1 is to the right of -2 on the number line and therefore, by the definition of inequality, $-1 > -2$.

Integers 47

Since 3 is to the right of -4 on the number line, $3 > -4$ or $-4 < 3$. Also note that $1 > 0$, $2 > 0$, and in general, all natural numbers are to the right of zero. Since all natural numbers, or positive integers, are to the right of zero, the inequality $n > 0$, for any integer n, is equivalent to the statement "n is a positive integer." Alternately, all negative integers are to the left of zero, so $n < 0$ is equivalent to the statement "n is a negative integer." Symbolically then, given an integer n:

If $n > 0$, then n is a positive integer or a natural number.
If $n < 0$, then n is a negative integer.

2.1. Exercises

Rewrite each of the following without using the absolute value symbols:

1. $|-12| =$
2. $|+7| =$
3. $|-0| =$
4. $|7 + 4| =$
5. $|-(9 + 6)| =$
6. $|12| + |8| =$
7. $|-3| + |-8| =$
8. $|32| + |-16| =$
9. $-(|-8|) = -|-8| =$
10. $-(|4|) = -|4| =$

Insert the proper symbol ($=$, $\neq$) between each of the following numbers:

11. $|-6|$ ____ $|+6|$
12. -8 ____ $+8$
13. $|7 + 4|$ ____ $|-7| + |-4|$
14. $+4$ ____ 4

Insert the proper symbol ($>$, $<$) between each of the following numbers:

15. -3 ____ 2
16. -4 ____ -3
17. 4 ____ -7
18. -81 ____ 2
19. $|-6|$ ____ 3
20. $|-8|$ ____ $|-17|$
21. -8 ____ -17
22. -5 ____ 0
23. -6 ____ $|+3|$
24. $|-(8 + 3)|$ ____ $-(|8 + 3|)$

State whether each of the following sentences are true or false:

25. $\{3,5\} \subseteq \{1,2,3,4\}$
26. $\emptyset \subseteq \{7,8,13\}$
27. $\{4,6,12\} \subseteq \{4,6,12\}$
28. $\{0,8,13\} \subseteq N$
29. $\{-1,-5,7\} \subseteq W$
30. $\{2,-3,0\} \subseteq J$
31. $\{|-3|,2,|+7|\} \subseteq N$
32. $\emptyset \subseteq \emptyset$

2.2. Addition of Integers

The number line was used in Chapter 1 as a means of visualizing the addition of two whole numbers. The number line can also be used here to suggest a method for the addition of two integers.

The addition, $3 + 2$, was performed in Chapter 1 by first starting at zero on the number line, then proceeding three units to the right to point A. From point A, an additional two units was marked to the right to point B, corresponding to 5. The graph of the procedure is shown in example (a). Now to perform the addition, $3 + (-2)$, proceed to point A as in example (a), then mark off two units to the left to point B. Point B corresponds to 1, shown in example (b) below. Note that the directed distance, -2, is indicated by a movement of two units to the left of point A.

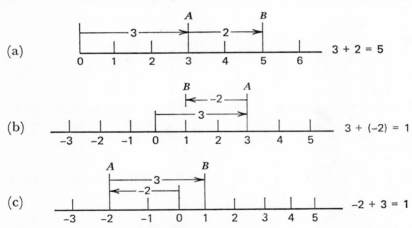

The parentheses which are inserted around the -2 in example (b) are intended to separate the "$+$" sign, which is the symbol for the operation of addition, and the "$-$" sign, which indicates a direction on the number line.

Again, to perform the addition, $-2 + 3$, proceed to point A which is two units to the left of zero. From point A, go three units to the right to point B, corresponding to 1. Refer to example (c) above. Now consider the addition, $-3 + (-2)$. Starting again at zero on the number line, proceed three units to the left to point A. From point A, go two units to the left to point B. Point B in example (d) below corresponds to −5. Finally, find the sum, −2 + (−3). From zero, proceed two units to the left to point A. From point A, proceed an additional three units to the left to point B. Again, point B in example (e) corresponds to −5.

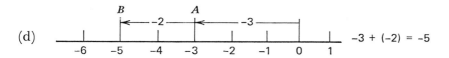

Examples (a), (b), (c), and (d) have illustrated the three combinations of adding two integers: two positive integers, two negative integers, and two integers having unlike direction signs.

Note that the sum of two positive integers shown in example (a) is found in the same way as the sum of two whole numbers. Examples (d) and (e) are both concerned with the addition of two negative integers. Referring to example (d), observe that first adding 3 and 2 and then finding the opposite of the sum, is equivalent to adding −3 and −2. This can be more clearly stated with symbols:

$$-3 + (-2) = -(3 + 2)$$
$$-2 + (-3) = -(2 + 3)$$

At the risk of being overly repetitious, these relations can now be generalized to give the statement:

> The sum of the negatives of two numbers is equal to the negative of their sum.

Before completing this section, a rather simple, but important concept can be developed with the following examples. A mental picture of the number line will aid in performing these operations:

(f) $2 + (-2) = 0$ and $-2 + 2 = 0$
(g) $-6 + (+6) = 0$ and $+6 + (-6) = 0$

50   Pre-Algebra

The sum of the examples (f) and (g) in either order add to 0, the identity element for addition.

> Whenever the sum of two numbers in either order produces the identity element, these numbers are said to be **additive inverses** of each other.

Specifically, -2 is the additive inverse of 2 because $-2 + 2 = 0$. Also, 2 is the additive inverse of -2 since $2 + (-2) = 0$. The additive inverse of a number then is found by taking the opposite, or negative of a given number. Since the additive inverse of 7 is -7, then the additive inverse of -7 is $-(-7)$. Now that the additive inverse of -7 is both 7 and $-(-7)$, then $-(-7) = 7$. This can be easily seen by referring to the number line and recalling that the negative of a number is in the opposite direction of the given number. Since -7 is seven units to the left of 0, the negative of -7, or $-(-7)$ is seven units to the right of 0. Now, 7 and $-(-7)$ are both seven units to the right of 0, so that $-(-7) = 7$.

The negative of the negative of a number is called the **double negative** of the number, so that:

> The double negative of a number is the given number. Stated symbolically:
> $$-(-b) = b \quad \text{where } b \text{ is any integer}$$

The addition of integers was intentionally left on an intuitive basis so that the student can develop a visual concept of addition of integers before more formal rules are developed in the next section.

2.2. Exercises

Find the sum of each of the following:

1. $+4 + (-2)$
2. $-4 + 9$
3. $7 + (-3)$
4. $-8 + 4$
5. $6 + (-8)$
6. $-7 + 3$
7. $9 + (-5)$
8. $-3 + 8$
9. $(+3) + (+5)$
10. $-9 + (-3)$

Integers 51

11. $7 + (+3)$
12. $(-12) + (-5)$
13. $-8 + 7$
14. $-6 + (-8)$
15. $-3 + (-7) + (-4)$
16. $-5 + 8 + (-4)$
17. $-3 + |-2|$
18. $8 + |-5|$
19. $|-5| + (-7)$
20. $13 + |-13|$
21. $|-7| + (-7)$
22. $|-9| + |-6|$
23. $|-4| + |-3| + |-2|$
24. $|+7| + |-9| + |+3|$
25. $-3 + 6 + (-7)$
26. $9 + (+6) + (-4)$
27. $-8 + (+7) + (-4)$
28. $(+8) + (-5) + (-7)$
29. $-4 + 8 + (-6)$
30. $-2 + (-5) + (-7)$
31. $5 + (-9) + (-8)$
32. $(-4) + (-9) + (-6)$

Simplify each of the following by removing all parentheses: (Note: [ ] (brackets) are used to eliminate confusion caused by the double use of parentheses.)

33. $-(+7)$
34. $-(-3)$
35. $+(+2)$
36. $+(-5)$
37. $-(|-4| + |-7|)$
38. $-[|-8| + (-4)]$
39. $-[9 + (-3)]$
40. $-[-7 + (-5)]$
41. $-(-8 + 2)$
42. $-[-(+7)]$
43. $-[+(-3)]$
44. $-[-(-4)]$
45. $-\{-[7 + (-12)]\}$
46. $-|-(1 - 31 + |-9|)|$
47. $-[4 + (-8) + 7]$
48. $-\{+[-(-13)]\}$

2.3. Addition Algorithm

The previous set of exercises should have provided the student with a clear, visual approach to the addition of integers. Unfortunately, this method is too slow to perform algebraic calculations efficiently and accurately. The following principle not only provides the foundation for an algorithm with which additions can be performed rapidly, but also shows that a relationship exists between the addition and subtraction operations.

In Section 2.1, the sum $3 + (-2)$ was found using the number line. Also note that $3 - 2 = 1$. Again, $8 + (-5) = 3$ as well as $8 - 5 = 3$.

Pre-Algebra

Observe that $3 - 2 = 3 + (-2)$ and $8 - 5 = 8 + (-5)$. Many more examples could be presented to substantiate the general principle that

$$a - b = a + (-b) \qquad \text{where } a \text{ and } b \text{ are any two numbers}$$

Since -2 is the additive inverse of 2, the equality $3 - 2 = 3 + (-2)$ can be stated: "Three minus two equals three plus the additive inverse of two." In general:

> The difference of two numbers is equal to the sum of the first and the additive inverse of the second.

Please note that the statement above defines the subtraction operation in terms of the addition operation. This new definition of subtraction demonstrates the similarity of the two operations.

Using the above definition, the addition of two integers having opposite signs can be performed with an equivalent subtraction problem. The following additions are done in this manner:

(a) $8 + (-3) = 8 - 3$ By the definition of subtraction
(b) $-4 + 7 = 7 + (-4)$ By the commutative law of addition
$ = 7 - 4$ By the definition of subtraction
$ = 3$

Now consider these examples:

(c) $4 + (-7) = 4 - 7$ By the definition of subtraction
(d) $-8 + 3 = 3 + (-8)$ By the commutative law of addition
$ = 3 - 8$ By the definition of subtraction

Examples (c) and (d) do not yield a solution as readily as examples (a) and (b). A short, simple method of performing addition with any two integers having opposite signs is obviously desirable. The following examples will aid in arriving at such an algorithm:

(e) $4 + (-7) = -3$ By the number line
$ = -(7 - 4)$ Since $3 = 7 - 4$
(f) $-8 + 3 = -5$ By the number line
$ = -(8 - 3)$ Since $5 = 8 - 3$

Observe in (e) that $4 = |4|$ and $7 = |-7|$, so that $4 + (-7) = -(|-7| - |4|)$. Again in (f), note that $8 = |-8|$ and $3 = |3|$, so that $3 + (-8) = -(|-8| - |3|)$. In both examples, the difference of the absolute values of both numbers were found. The sign of the difference is negative, which is the sign of the number having the largest absolute value. Consider two more examples:

(g) $6 + (-2) = +(6 - 2)$ By the definition of subtraction
(h) $-5 + 7 = 2$ By the number line
$ = +(7 - 5)$ Since $2 = 7 - 5$

Observe in (g), that $6 = |6|$ and $2 = |-2|$, so that $6 + (-2) = +(|6| - |-2|)$. In (h), note that $5 = |-5|$ and $7 = |7|$, so that $-5 + 7 = +(|7| - |-5|)$. Here again, both examples obey the same rule as above except that the sign of the difference is positive, which is the sign of the number having the largest absolute value. Restating this in a more formal manner:

> When adding two integers having unlike signs, find the difference of their absolute values. The sign of the difference is the sign of the number having the largest absolute value.

The addition of two negative integers can also be defined in terms of their absolute values, as shown in the next examples:

(i) $-3 + (-2) = -(3 + 2)$ Definition of the sum of two negative numbers
$ = -(|-3| + |-2|)$ Since $3 = |-3|$ and $2 = |-2|$

(j) $-6 + (-7) = -(6 + 7)$ Definition of the sum of two negative numbers
$ = -(|-6| + |-7|)$ Since $6 = |-6|$ and $7 = |-7|$

Note that:

> The sum of two negative integers is equal to the negative of the sum of their absolute values.

54   Pre-Algebra

Finally, since $4 = |4|$ and $3 = |3|$, $4 + 3 = |4| + |3|$. It should be clear without additional discussion that:

> The sum of two positive integers is equal to the sum of their absolute values.

By the use of numerous examples, algorithms have been devised for the three cases of adding two integers. Summarizing:

1. The sum of two positive integers is equal to the sum of their absolute values.
2. The sum of two negative integers is equal to the negative of the sum of their absolute values.
3. The sum of two integers having unlike signs is equal to the difference of their absolute values. The sign of the difference is the same as the sign of that number having the largest absolute value.

In conclusion, algorithms were developed in this section to add any two integers. By using numerical examples, the student should confirm that:

> The set of integers is commutative with respect to addition. The set of integers is associative with respect to addition. The definition of inequality developed for whole numbers in Section 1.7, p. 34, also applies to integers. The identity element for addition developed for whole numbers in Section 1.3, p. 16, also holds for integers.

It remains to be seen whether or not the above properties hold for subtraction of integers. These questions shall be explored in Section 2.4.

2.3.   Exercises

Simplify each of the following expressions to a single integer. Remove all parentheses and perform all operations. Use the algorithms wherever possible:

1. $8 + (-11)$
2. $-5 + (-6)$

Integers 55

3. $9 + (-4)$
4. $-7 + (-6)$
5. $-9 + (+6)$
6. $-7 + 4$
7. $-5 + 13$
8. $-3 + (+12)$
9. $+6 + (+13)$
10. $-8 + (-3)$
11. $(+8) + 7$
12. $-9 + (+13)$
13. $(-8) + (-3) + (-7)$
14. $(-4) + 7 + (-5)$
15. $12 + (-7) + (-8)$
16. $(-11) + 5 + (-4)$
17. $7 - 4 + (-6)$
18. $12 + (-5) - 4$
19. $(-8) + (-5) + 17$
20. $(-3) + (-6) + (-7)$
21. $-6 + (-8) + 12$
22. $(14 - 8) - 4$
23. $-14 - (12 - 7)$
24. $-[6 + (-9)] + (-8)$
25. $14 - (8 - 4)$
26. $[12 + (-6)] + (-2)$
27. $-[8 + (-3)] + (-7)$
28. $-[(-6) + 13 + (-9)]$
29. $12 + [(-6) + (-2)]$
30. $8 + (-9) + (-3) + 6 + (-5)$
31. $-[7 + (-12)] + [-(3 + 9)]$
32. $(-7) + 4 + (-5) + (-6) + 9$
33. $436 + (-875) + (-87) + 123$
34. $-98 + (-137) + (-923) + (-2,046) + 1,327$
35. $27 + (-83) + (-235) + (-316)$
36. $127 + (-81) + (-216) + 618 + (-513)$

56   Pre-Algebra

2.4. Subtraction of Integers

In order to develop the desired algorithms for addition of integers in the previous section, it was necessary to define subtraction. Referring to that section, it can be seen that only the differences of positive integers were found when using the addition algorithm. In this section, subtraction of any two integers shall be developed in two cases: subtraction of a positive integer from any given integer, and subtraction of a negative integer from any given integer:

(a) $(+8) - 3 = 8 + (-3)$ By definition of subtraction
$= 5$ By addition algorithm or number line

(b) $-6 - (+3) = -6 + (-3)$ By definition of subtraction
$= -9$ By addition algorithm or number line

By converting these subtraction problems into addition problems, solutions can then be found by using the addition algorithm or the number line. The next examples illustrate subtraction of a negative integer from any given integer:

(c) $+8 - (-2) = 8 + [-(-2)]$ By definition of subtraction
$= 8 + 2$ The double negative of a number
$= 10$

(d) $-5 - (-3) = -5 + [-(-3)]$ By definition of subtraction
$= -5 + 3$ The double negative of a number
$= -2$ By addition algorithm or number line

The previous examples illustrated all the possible situations which may arise when subtracting two integers. In each case the subtraction problem was replaced by an equivalent addition problem and the addition algorithm was used. When subtracting a negative number, the double negative of the number is replaced by the positive of the number, and then added.

Therefore, in all cases,

$$a - b = a + (-b) \quad \text{where } a \text{ and } b \text{ are any two integers}$$

The question continually arises as to whether the "$-$" sign encountered in any expression is a minus sign or a negative sign. The answer depends upon how the statement is written. Consider these statements:

(e) $\quad 7 - 2 = 7 - (+2)$
(f) $\quad 7 - 2 = 7 + (-2)$

In (e), the "$-$" is a minus sign, subtracting a positive 2. In (f), the "$-$" is converted into a negative sign, adding a negative 2. Because of the definition of subtraction given in Section 2.3, p. 52, the "$-$" can be a minus sign or a negative sign, depending upon the interpretation desired. A subtraction problem then can be treated as in (e) or (f), whichever is most convenient.

The "$+$" sign is also used in two ways without confusion. For example, $3 + 8$ can be written as $3 + (+8)$. The first "$+$" is the addition operation sign. The second "$+$" represents a positive sign, or the direction of 8 on the number line. The double positive is unnecessary and is seldom used.

A convention was adopted on Section 1.5, p. 26, concerning mixed operations involving additions and/or subtractions of whole numbers. This convention applies to integers as well, illustrated by the following examples:

(g) $\quad 8 - 3 + 5 = 5 + 5$
$ = 10$

(h) $\quad -7 + 8 - 5 - 7 = 1 - 5 - 7 \qquad\qquad$ Since $-7 + 8 = 1$
$ = -4 - 7 \qquad\qquad$ Since $1 - 5 = -4$
$ = -11$

(i) $\quad 4 - 7 - (-5) - 3 = -3 - (-5) - 3 \qquad\qquad$ Since $4 - 7 = -3$
$ = -3 + 5 - 3 \qquad\qquad$ Since $-(-5) = 5$
$ = 2 - 3 \qquad\qquad$ Since $-3 + 5 = 2$
$ = -1$

Restating the rule at this time:

> Mixed operations, involving only additions and/or subtractions, are performed from left to right, unless otherwise indicated by grouping symbols such as parentheses.

Since $6 - 4 \neq 4 - 6$, subtraction is not a commutative operation, as stated in Section 1.5, p. 26. This fact applies to integers as well. In fact, it can be generally stated that,

$$a - b \neq b - a \quad \text{for all integers, providing that } a \neq b$$

Again then, subtraction is not generally a commutative operation.

Observe that starting with 0, first adding 7, and then subtracting 7, results in 0 again. Starting with -6, first adding 4, and then subtracting 4, results in -6 again. In fact, starting with any integer b, adding any integer a, and then subtracting a, results in the integer b again. This is also true if a is first subtracted, and then added. That is, $b + a - a = b$ and $b - a + a = b$. Since subtraction "undoes" the effect of addition, and vice versa, they are referred to as **inverse operations.** Starting with any element, if any combination of two operations results in the given element again, these operations are called inverse operations. Specifically, addition is the inverse operation of subtraction, subtraction is the inverse operation of addition, and it can be said that subtraction and addition are inverse operations.

It was stated in Section 1.3, p. 16, that 0 is the identity element for addition provided that its sum with any number in **either order** did not change the identity of the given element. Now observe that $7 - 0 = 7$, but $0 - 7 \neq 7$. Many more examples could be provided to show that 0 cannot be subtracted from a nonzero number in either order and produce the same result. Therefore, 0 is not the identity element of subtraction and it can be shown that there is no identity element for subtraction.

Finally, $(8 - 3) - 5 = 0$ and $8 - (3 - 5) = 10$, so it should be clear that $(8 - 3) - 5 \neq 8 - (3 - 5)$. Since many examples of this type can be demonstrated, it can be stated generally that:

> Subtraction of integers is **not** an associative operation.

2.4. Exercises

Perform each of the following subtraction problems:

1. $6 - 11$
2. $-9 - 6$
3. $-8 - 5$
4. $-2 - 9$
5. $+8 - (-3)$
6. $+3 - |-4|$
7. $7 - (+3)$
8. $8 - |-5|$
9. $-5 - (-8)$
10. $7 - (-5)$
11. $-12 - 14$
12. $13 - (-4)$
13. $2 - (-7)$
14. $-8 - (-7)$
15. $-9 - (-4)$
16. $-4 - (-9)$
17. $|-4| - |-8|$
18. $|-7| - |-3|$
19. $|-6| - 3$
20. $|-2| - |+7|$
21. $+2 - |+7|$
22. $+9 - |+4|$
23. $12 - (+3)$
24. $+387 - 816$

Find a single integer equal to each of the following expressions:

25. $7 - 3 - 8$
26. $-3 + (-5)$
27. $-5 + 8 - 9$
28. $+5 - 6 - 7$
29. $-8 - 5 - 7$
30. $(-2) - 5 + 3$
31. $-7 + 4 + 9$
32. $-7 - (-5) - (-8)$
33. $2 + 8 - 3$
34. $7 - (-4) - |-2|$
35. $2 - (-7) - 8$
36. $|-3| - |-7| + |-5|$
37. $6 - 8 - (-7)$
38. $-4 + (-5) - (-2)$
39. $-(-1) + 5 - (-8)$
40. $8 + (-3) - (-5)$
41. $-4 + (-6) - (-3)$

Pre-Algebra

Find a single integer equal to each of the following expressions:
42. $|+5| + (-3) + (-5)$
43. $-(-3) - 4 + (-6)$
44. $6 - (-3) - 4 + (-2) + 7$
45. $-3 - 7 + (-4)$
46. $9 + 4 + (-9) - 5 - (-3) + 8$
47. $-285 + 483$
48. $7 - 9 + 3 - 8 - 3 + 5 - (-7)$
49. $193 - 817 + 436$
50. $-217 + (-48) - (-1{,}235) + 813$
51. $-4{,}026 - (-394) - 1{,}817$
52. $3 - 8 + 2 - (-7) - 6 + 4$
53. $-(12 + 5) - |-(8 - 2)| - [-(7 + 5)]$
54. $|-(8 + 3)| - (9 + 5) - [-(6 - 3)] + [-(7 - 4)]$

Supply the missing symbol ($>$, $=$, $<$) between each of the following:
55. $6 \underline{\hspace{1em}} -(-6)$
56. $-2 \underline{\hspace{1em}} -6$
57. $+8 \underline{\hspace{1em}} -(-4)$
58. $-3 - 4 \underline{\hspace{1em}} -3 + (-4)$
59. $6 - 9 \underline{\hspace{1em}} 9 - 6$
60. $|-2| \underline{\hspace{1em}} |-6|$
61. $7 - (-8) \underline{\hspace{1em}} 8 + 7$
62. $|9 - 13| \underline{\hspace{1em}} |13 - 9|$
63. $|5 - 7| \underline{\hspace{1em}} |7 - 5|$
64. $-(8 - 3) \underline{\hspace{1em}} 3 - 8$
65. $(8 - 6) - 3 \underline{\hspace{1em}} 8 - (6 - 3)$
66. $-|-7| \underline{\hspace{1em}} -(-7)$
67. $|-2 - 8| \underline{\hspace{1em}} 2 + 8$
68. $(-2 - 4) - 6 \underline{\hspace{1em}} -2 - (4 - 6)$
69. $8 - 11 \underline{\hspace{1em}} 11 - 8$
70. $-[-(4 - 9)] \underline{\hspace{1em}} 9 - 4$
71. $[-(3 - 8)] - 5 \underline{\hspace{1em}} 5 - (8 - 3)$
72. $(9 + 5) - 6 \underline{\hspace{1em}} 9 + (5 - 6)$

2.5. Multiplication of Integers

The multiplication of whole numbers was defined in Section 1.6 and it was also established there that

$$a \cdot b = b \cdot a$$
$$a \cdot (b \cdot c) = (a \cdot b) \cdot c \qquad \text{where } a, b, \text{ and } c \text{ are any whole numbers}$$

The first statement is called the commutative property of multiplication, while the second is called the associative property of multiplication.

There are many ways to establish that the multiplication of integers is commutative and associative. For the purposes of this discussion, it will suffice to assume these properties. Henceforth, then,

$$a \cdot b = b \cdot a$$
$$a \cdot (b \cdot c) = (a \cdot b) \cdot c \qquad \text{where } a, b, \text{ and } c \text{ are any integers}$$

Now for the specific cases of multiplication of integers. Observe that:

$$(-1) + (-1) = 2(-1)$$

also

$$(-1) + (-1) = -2$$

so that

$$2(-1) = -2$$

Using the property that multiplication of integers is commutative, $2(-1) = (-1) \cdot 2$ and therefore, $-1 \cdot 2 = -2$. Since $-1 \cdot 2$ results in the opposite, or negative of 2, it is reasonable to assume that $-1 \cdot (-3)$ produces the opposite of -3, or $-(-3)$. Therefore, $-1 \cdot 2 = -2$ and $-1 \cdot (-3) = -(-3)$, or more generally,

$$-1 \cdot a = -a \qquad \text{for any integer } a$$

Using this new relationship,

(a) $\quad -3 \cdot 2 = (-1 \cdot 3) \cdot 2 \qquad\qquad -3 = -1 \cdot 3$
$\qquad\quad\; = -1 \cdot (3 \cdot 2) \qquad\qquad$ Associative property of multiplication
$\qquad\quad\; = -1 \cdot 6 \qquad\qquad\qquad 3 \cdot 2 = 6$
$\quad -3 \cdot 2 = -6 \qquad\qquad\qquad\quad\; -1 \cdot 6 = -6$

Pre-Algebra

Again,

(b) $2(-3) = 2(-1 \cdot 3)$ $-3 = -1 \cdot 3$
$ = [2 \cdot (-1)]3$ Associative property of multiplication
$ = (-1 \cdot 2) \cdot 3$ Commutative property of multiplication
$ = -1 \cdot (2 \cdot 3)$ Associative property of multiplication
$ = -1 \cdot 6$
$2(-3) = -6$ $-1 \cdot a = -a$

In a like manner it can be shown that, $4(-6) = -24$, $-5(3) = -15$, $8(-4) = -32$, and $-7(2) = -14$. Observe that in all cases, a positive number multiplied by a negative number gave a negative product. Examples (a) and (b) can also be interpreted as follows:

$$-3(2) = -(|-3| \cdot |2|)$$

$$2(-3) = -(|2| \cdot |-3|)$$

This can be generalized by the statement:

> The product of two integers with unlike signs is equal to the negative of the product of their absolute values.

Representing a and b as any two natural numbers, this becomes

$a(-b) = -(|a| \cdot |-b|) = -(a \cdot b)$
$(-a) \cdot b = -(|-a| \cdot |b|) = -(a \cdot b)$

Now consider the following examples of multiplication of two negative integers:

(c) $-3 \cdot (-4) = [(-1) \cdot 3] \cdot (-4)$ Since $-a = -1 \cdot a$
$ = (-1)[3 \cdot (-4)]$ Associative property of multiplication
$ = -1 \cdot (-12)$ $3(-4) = -12$
$ = -(-12)$ Since $-1 \cdot a = -a$
$-3(-4) = 12$ The double negative property

(d) $(-6) \cdot (-7) = [(-1) \cdot 6] \cdot (-7)$ Since $-1 \cdot a = -a$
$ = (-1) \cdot [6(-7)]$ Associative property of multiplication
$ = -1 \cdot (-42)$ $6(-7) = -42$
$ = -(-42)$ Since $-1 \cdot a = -a$
$ = 42$ The double negative property

Additional examples could be shown to support the conclusion that:

The product of two negative integers is equal to the product of their absolute values.

Rewriting this statement symbolically:

$(-a) \cdot (-b) = |-a| \cdot |-b| = a \cdot b$ where a and b are any two natural numbers

The zero factor property, $a \cdot 0 = 0 \cdot a = 0$, developed for whole numbers in Section 1.6, p. 31, can be extended to integers. This can be seen by the following example:

(e) $0(-7) = -7 \cdot 0$ Commutative property of multiplication
$ = -(7 \cdot 0)$ $-a \cdot b = -(a \cdot b)$
$ = -(0)$ $a \cdot 0 = 0$
$0(-7) = 0$ $-0 = +0 = 0$

Therefore, $0(-7) = -7 \cdot 0 = 0$, and these results can be easily generalized so that,

$a \cdot 0 = 0 \cdot a = 0$ for any integer a

That is, the zero factor property is valid for all integers. Specifically, the product of zero and any integer is zero.

Since the product of two positive integers needs no additional discussion, the product of any two integers can be summarized as follows:

1. The product of any two positive integers is equal to the product of their absolute values.
2. The product of any two negative integers is equal to the product of their absolute values.

64   Pre-Algebra

3. The product of a positive integer and a negative integer is equal to the negative of the product of their absolute values.
4. The product of zero and any integer equals zero.

In Chapter 1, the identity element of multiplication was found to be 1. Considering that $-3 \cdot 1 = -3$ and $1 \cdot (-3) = -3$, it should be clear that the identity of any given integer is unchanged when multiplied by 1 in either order. Therefore, for integers:

> 1 is the identity element for multiplication

Calculations involving mixed operations with whole numbers were also discussed in Chapter 1. The convention for mixed operations which was stated there also applies to integers and is repeated below as a reference:

> When multiplication, along with addition and/or subtraction, occurs in any given expression, all multiplications are performed first, providing that grouping symbols such as parentheses not not indicate otherwise. The remaining additions and subtractions are then performed, proceeding from left to right.

The following examples which include mixed operations illustrate this convention:

(f) $\quad -4 \cdot 3 + 6 = -12 + 6$
$\qquad\qquad\qquad\quad = -6$

(g) $\quad 7 - 2 \cdot 8 = 7 - 16$
$\qquad\qquad\qquad = -9$

(h) $\quad -4(-6) - 3 = 24 - 3$
$\qquad\qquad\qquad\quad = 21$

In conclusion, algorithms were derived in this section for the multiplication of two integers. Multiplication of integers is both commutative and associative. That is:

$$a \cdot b = b \cdot a$$
$$a \cdot (b \cdot c) = (a \cdot b) \cdot c$$
$$a(-b) = -(a \cdot b)$$
$$-a \cdot b = -(a \cdot b)$$
$$(-a)(-b) = a \cdot b$$

where a, b, and c are any integers

As well as:

$$-1 \cdot a = -a$$
$$a \cdot 0 = 0 \cdot a = 0 \quad \text{for any integer } a$$

2.5. Exercises

Perform each of the multiplications as indicated. Express your answer as a single, simplified integer:

1. $7(-4)$
2. $-6(-8)$
3. $-3(8)$
4. $5(-7)$
5. $(+3)(+6)$
6. $-8(+3)$
7. $-[(-4)(-9)]$
8. $9(+8)$
9. $(-5)(+3)(-2)$
10. $9(-3)(+2)$
11. $4(-6)(-3)$
12. $-8(2)(-4)$
13. $(-4)(-5)(-3)$
14. $(-8)(3)(-3)$
15. $-[3(-6)(-5)]$
16. $-[(-7)(-2)(-3)]$
17. $3(9-2)$
18. $-5(8+3)$
19. $-7[3+(-8)]$
20. $4(6-8)$
21. $-3(7-9)$
22. $(9-6)(7-12)$
23. $-(3-8)(7-12)$
24. $-3(8-2)(3-9)$

Perform each of the following mixed operations as indicated. Express your answer as a single, simplified integer:

25. $3(-5) + 8$
26. $(-4)(-5) - 3$
27. $-6 + 5(-3)$
28. $-8 - 4(-7)$
29. $9(-3) + (-4)(-7)$
30. $(-8)(-3) - (4)(-4)$
31. $5(-8) + 7(-4)$
32. $6(-5) - (-4)(8)$
33. $7 + 3 \cdot (-4) - 5$
34. $-3 - 8 + 6 \cdot (-3)$
35. $8 - 4(-6) - 7$
36. $(-5) \cdot 8 - (-3)(-7)$
37. $7(-3) + 6 - 5(-2)$
38. $4(-5) - (-7)(+3) + 9$
39. $-3 - 6 \cdot 7 - 8(-6) - 3$
40. $-7(8-5) - 3(7-12)$

Supply the missing symbol ($<$, $=$, $>$) between each of the following:

41. $-3 \cdot |-4|$ _____ $-3 \cdot (-4)$
42. $(6{,}837)(4{,}938)$ _____ $(-836)(23{,}916)$

Pre-Algebra

43. $(743)(916)$ _____ $(-743)(-916)$
44. $(3,937)(61,834)$ _____ $(3,938)(61,835)$
45. $-3(6-9)$ _____ $3(9-6)$
46. $47 \cdot 0$ _____ $-3(8-7)$
47. $|(-81)(-57)(+38)|$ _____ $|(81)(-57)(-38)|$
48. $|(237)(-184)(-385)|$ _____ $385 \cdot |(-237)(-184)|$

2.6. Distributive Property

Multiplication and addition have been defined in previous sections as two separate operations. The **distributive property,** defined in this section, shall provide a connecting link between the two operations. Observe that the pattern emerging from the following examples serves as motivation for this new property:

(a) $\quad 6(2+3) = 30$
$\qquad\qquad\quad = 12 + 18$
$\quad\; 6(2+3) = 6 \cdot 2 + 6 \cdot 3$

(b) $\;\; -4[7+(-2)] = -20$
$\qquad\qquad\qquad = -28 + 8$
$\;\; -4[7+(-2)] = -4 \cdot 7 + (-4)(-2)$

(c) $\quad -6[-5+(-2)] = 42$
$\qquad\qquad\qquad = 30 + 12$
$\;\; -6[-5+(-2)] = -6 \cdot (-5) + (-6)(-2)$

(d) $\;\; +6[4+(-8)] = -24$
$\qquad\qquad\qquad = 24 + (-48)$
$\;\; +6[4+(-8)] = 6(4) + 6(-8)$

Note that in each example, the integer outside the parentheses was **distributed** over the sum of the two integers inside the parentheses. This property is simply stated:

Multiplication can be distributed over addition.

This is referred to as the distributive property. Since multiplication is a commutative operation, example (a),

$$6(2+3) = 6 \cdot 2 + 6 \cdot 3,$$

can be restated as:
$$(2 + 3) \cdot 6 = 2 \cdot 6 + 3 \cdot 6$$

Generalizing these properties symbolically:

$a \cdot (b + c) = a \cdot b + a \cdot c$
$(b + c) \cdot a = b \cdot a + c \cdot a$ for a, b, and c any three integers

The distributive property shall prove to be one of the most useful and important relationships in algebra.

Referring to example (d),
$$6[4 + (-8)] = 6 \cdot 4 + 6(-8)$$

Note that both sides of the example can be restated as a subtraction problem as follows:

$$6(4 - 8) = 6 \cdot 4 + (-48)$$
$$= 6 \cdot 4 - 48$$
$$6(4 - 8) = 6 \cdot 4 - 6 \cdot 8$$

Looking again at example (b),
$$-4[7 + (-2)] = -20$$

This can be restated as:

$$-4(7 - 2) = -4(7) - (-4)2$$
$$= -4(7) - (-8)$$
$$= -28 + 8$$
$$= -20$$

The result is the same as if the left side was treated as a sum. The examples should make it obvious that:

Multiplication can be distributed over subtraction.

Symbolically, this becomes:

$a(b - c) = a \cdot b - a \cdot c$
$(b - c) \cdot a = b \cdot a - c \cdot a$ for a, b, and c any three integers

Occasionally, the distributive property is very useful when used in reverse. Before demonstrating this result, it will first be necessary to introduce some additional terms. When numbers are multiplied together to produce a product, each number is called a **factor** of that product. The process of finding these factors is called **factoring.** Thus, 3 and 4 are factors of 12; 6 and 2 are factors of 12; 3 and 5 are factors of 15; 2 and 14 are factors of 28, and 4 and 7 are factors of 28. Notice that the set of factors of a given number is not necessarily unique. That is, some numbers have many sets of factors. When two numbers have a factor in common, this factor is called a **common factor.** For example, 15 and 6 have a common factor of 3; 48 and 12 have common factors of 2, 3, 4, 6, and 12. Every pair of numbers has a factor of 1 in common. The largest factor which two numbers have in common is called the **greatest common factor,** sometimes abbreviated as the **GCF.**

The following examples demonstrate how the distributive property is related to factoring:

(e) $\quad 6 \cdot 27 + 4 \cdot 27 = (6 + 4) \cdot 27$
$\qquad\qquad\qquad\quad= 10 \cdot 27$
$\qquad\qquad\qquad\quad= 270$

(f) $\quad -7 \cdot 43 + 5 \cdot 43 = (-7 + 5) \cdot 43$
$\qquad\qquad\qquad\qquad= -2 \cdot 43$
$\qquad\qquad\qquad\qquad= -86$

(g) $\quad 87 \cdot 3 + 87 \cdot (-2) = 87 \cdot [3 + (-2)]$
$\qquad\qquad\qquad\qquad\;= 87 \cdot 1$
$\qquad\qquad\qquad\qquad\;= 87$

In the three previous examples, the distributive property was used as originally defined, but as follows:

$$a \cdot b + a \cdot c = a \cdot (b + c)$$

and

$$b \cdot a + c \cdot a = (b + c) \cdot a$$

In example (e), 6 and 27 are factors of the product of 6 and 27. Likewise, 4 and 27 are factors of the product of 4 and 27. The factor 27 is a common factor of the two products, since it occurs in both. Observe in examples (e), (f), and (g) that the distributive property permits the sum of the products to be rewritten as the product of the common factor with the sum of the remaining factors. With this result, the calculations for these examples require one addition and one multiplication rather than the two multiplications and one addition required in the original expressions. The factoring process, that of removing common factors as in the examples above, is frequently useful for simplifying calculations. This maneuver becomes very effective in algebra and other advanced mathematics courses.

2.6. Exercises

In each of the following problems, first use the distributive property to distribute multiplication over addition, and then add or subtract as indicated to simplify to a single integer. Check your results by first adding or subtracting as indicated, and then multiplying:

Example:
$-3[6 + (-4)] = -3 \cdot 6 + (-3)(-4)$
$= -18 + 12$
$= -6$

Checking:
$-3[6 + (-4)] = -3 \cdot 2$
$= -6$

1. $4[4 + (-7)]$
2. $-6(8 + 3)$
3. $7[(-6) + 4]$
4. $-6(9 - 6)$
5. $-8[7 - (-2)]$
6. $-3[-5 - (-3)]$
7. $9[-4 - (-5)]$
8. $5[+6 - (-3)]$
9. $-3[-3 + (-8)]$
10. $-4(6 - 9)$
11. $-8[-7 - (-3)]$
12. $6[8 - (-4)]$

Replace each of the following expressions with a single integer:

13. $4(-3) + 5(-6 + 2)$
14. $-4 \cdot (-7 - 3) - 6(-3)$
15. $4 |-3| - 7 |6 - (-2)| + |-3| \cdot |-6|$
16. $(8 - 11) \cdot (11 - 8) - 3(-3)$
17. $-6(-2)(-4)$
18. $4,376 \cdot 0 + 8$

19. $17(4 - 8 + 4) - 83(-17 + 23 - 6)$
20. $-8(-4 + 7) + 5[7 - (-9)]$
21. $-3 + 8(7 - 9)$
22. $(-4 - 8) \cdot 3 + 6$
23. $-5 |8 - 3| - 3(8 - 5) + (-6) \cdot 7$
24. $(9 - 7)(-6 - 7) - (4 - 7)(7 - 13)$

Perform each of the following calculations by using the distributive property to write them as problems requiring one multiplication:

25. $43 \cdot 8 + 43 \cdot 2$
26. $7 \cdot 16 + 3 \cdot 16$
27. $29(13) - 29(-3)$
28. $23 \cdot 17 + (-3) \cdot 17$
29. $-4 \cdot 116 + (-6) \cdot 116$
30. $5(34) + 3(34) + 2(34)$
31. $8 \cdot (-12) - 13(-12)$
32. $-7 \cdot 16 + 13(-7)$
33. $6(-18) + (-18)(-14)$
34. $(-6)(-3) \cdot 9 + (7)(-3) \cdot 9$
35. $5 \cdot (-8) \cdot 9 - 4 \cdot (-8) \cdot (7)$
36. $12(-8)(-6) + (-17)(-8)(-6)$

Write all the sets of factors for each of the following numbers:

37. 18 38. 40 39. 42 40. 56

State the greatest common factor (GCF) for each of the following pairs of numbers:

41. 6, 9
42. −4, −12
43. 7, 9
44. 12, 18
45. 10, 25
46. −8, 28
47. 16, 18
48. 42, 49
49. 26, 39
50. 126, 270

2.7. Summary

In this chapter, the set of whole numbers was extended to the set of integers. This was motivated by the desire to have a set of numbers which

Integers 71

was closed with respect to subtraction. The commutative and associative properties, and the identity element for addition, developed for whole numbers, were also extended to the set of integers. Subtraction and multiplication were defined, and algorithms for these operations were constructed for the set of integers. The commutative and associative properties, and the identity element for multiplication were defined for the extended set. The distributive property was found to be a unifying concept relating multiplication and addition, as well as multiplication and subtraction.

The ordering properties developed for whole numbers were extended to the set of integers. The zero factor property was also defined in this chapter, as well as the double negative property and the concept of additive inverse. The absolute value of a number was introduced, and algorithms for all operations involving integers were stated in terms of this new concept.

Finally, some guidelines were set down for dealing with mixed operations involving addition, subtraction, and multiplication.

2.7. Review Exercises

Simplify each of the following expressions. The final result should be in the form of a single integer expressed without any absolute value signs:

1. $3 + (-6)$
2. $-7 + (-6)$
3. $-(-4) + |-4|$
4. $-5 - 8$
5. $-|4 + (-7)|$
6. $-5 + |-5|$
7. $6 - (-3) - 7$
8. $-4 + 7 - 12$
9. $7 - 3(8 - 11) + 3(-5)$
10. $9 - 5 \cdot 3 + 7$
11. $-3 - 8 - 6$
12. $-381 + 783 - 402$
13. $8 - (-3) + (-6) - 4 + 2 - 6$
14. $+218 + (-617) - (+438) - (-2{,}134)$

Pre-Algebra

15. $-3 \cdot 2 + (-6)(-4)$
16. $-4(-7)(-2)$
17. $(+3)(-5) - (7)(-6)$
18. $-[(-3)(-9) - 7]$
19. $-4 + 6(-7) + 13$
20. $-[-(-8)(-4)]$
21. $(417)(-32)(+163) \cdot 0$
22. $4(-6) - 8 - 6(-3)$
23. $(4 - 9)[7 - (-3)]$
24. $-3[4 + (-6)(3) + 9]$

Supply the missing symbol ($>$, $=$, $<$) between each of the following:

25. $|-6| \cdot |+3|$ ____ $-6 \cdot (+3)$
26. -3 ____ -2
27. $|8 - 11|$ ____ $11 - 8$
28. $-|7 + 6|$ ____ $-(7 + 6)$
29. $|-5 + 8|$ ____ $|-5| + |8|$
30. $|4 - (-6)|$ ____ $|4| - |-6|$
31. $4 - (7 - 9)$ ____ $(4 - 7) - 9$
32. $8 \cdot 4 - 9$ ____ $8(4 - 9)$
33. $-3(3 - 8)$ ____ $-3 \cdot 3 - (-3)(8)$
34. $-[-(2 - 7)]$ ____ $-(7 - 2)$
35. $218(-693)$ ____ $318(476)$
36. $|(-4,837)[418 - 346]|$ ____ 0

Write all the factor sets for each of the following:

37. 24   38. 36   39. 54   40. 63

Find the greatest common factor for each of the following pairs of numbers:

41. 8, 12   42. -9, 16   43. 18, 30   44. 45, 105

Complete the following statements:

45. The additive inverse of -3 is _____.
46. The double negative of -6 is _____.
47. Since $(6 - 9) + 3 \neq (9 - 6) + 3$, subtraction is not a(n) _____ operation.
48. Multiplication is defined as _____ _____.

Circle the correct answer:

49. The absolute value of a number is ($>$, $\geq$, $=$, $<$, $\leq$) zero.

50. The absolute value of a number and the number are (never, sometimes, always) equal.
51. The absolute value of the sum of two integers is (always, sometimes, never) equal to the sum of the absolute values of these two integers.
52. When subtracting a negative number, proceed to the (left, right) on the number line.

3 Rational Numbers

3.1. Multiplication of Rational Numbers

In developing the number system thus far, an attempt has been made to justify each new extension or property before it is presented. By considering some questions, it should again become obvious that an extension of the set of integers is necessary in order to be able to express the answers to these questions:

(a) How much of the pie has Bob eaten if it was cut into five pieces and he then consumed three of these?
(b) If Lori's horse eats one bale of hay in four days, how much hay should she set out each day in order to keep the horse's diet the same?
(c) If Mark is to walk two miles in three hours, how many miles must he walk each hour in order to walk the same distance each hour?

In these and many more instances, the set of integers cannot express the answers to the questions and, therefore, new numbers are needed.

Pre-Algebra

In example (a), Bob has eaten 3 pieces out of a total of 5 pieces. This can be written as $\frac{3}{5}$. In example (b), Lori's horse eats 1 bale of hay in a total of 4 days, or $\frac{1}{4}$ of a bale each day. In example (c), Mark will walk 2 miles in 3 hours, or $\frac{2}{3}$ of a mile each hour. These are read as three-fifths, one-quarter, and two-thirds in examples (a), (b), and (c), respectively. Each of these numbers is called a **rational number,** and generally:

> Rational numbers are those numbers which can be written in the form $\frac{a}{b}$, read as "a over b," where a and b are integers, $b \neq 0$

If $b = 1$, a particular set of rational numbers is obtained, which is of the form, $\frac{a}{1}$. Specifically, if an object measures 7 units, each unit being 1 foot long, then the object is 7 feet long. If 3 eggs are to be put into groups each of which contains one egg, then there will be 3 groups. Therefore, $\frac{7}{1} = 7$ and $\frac{3}{1} = 3$, so that, in general:

$$\frac{a}{1} = a \quad \text{where } a \text{ is any integer}$$

Therefore, rational numbers which are of the form, $\frac{a}{1}$, are actually integers, which means that every integer is also a rational number.

Letting Q represent the set of rational numbers, and recalling that

$N = \{\text{natural numbers}\} \quad W = \{\text{whole numbers}\} \quad J = \{\text{integers}\}$

then the following relations hold:

$$J \subseteq Q \quad N \subseteq J \quad N \subseteq Q \quad W \subseteq Q$$

Rational numbers have an interpretation on the number line. If 1 is considered to be the whole pie in example (a), then $\frac{1}{5}$ represents 1 piece of the 5 pieces of pie. On the number line, a length of 1 unit can be subdivided into 5 equal lengths as indicated below:

(d)

$$\underset{0 \quad\quad \frac{1}{5} \quad\quad \frac{2}{5} \quad\quad \frac{3}{5} \quad\quad \frac{4}{5} \quad\quad 1}{\longleftarrow \tfrac{1}{5} \to \leftarrow \tfrac{1}{5} \to \leftarrow \tfrac{1}{5} \to \leftarrow \tfrac{1}{5} \to \leftarrow \tfrac{1}{5} \to}$$

Rational Numbers 77

Here, 2 pieces of pie represent two parts of the five parts. This can be represented as 2 parts of 5 parts equaling $\frac{2}{5}$, or two $\frac{1}{5}$'s (read as two one-fifths) and written as $2(\frac{1}{5})$. Since both represent the same quantity of pie, $2(\frac{1}{5}) = \frac{2}{5}$.

Again consider example (b). The bale of hay can be considered as 1 unit (one bale), and separated into three parts as shown below:

(e)

$$\overset{\left|\leftarrow\frac{1}{3}\rightarrow\right|\leftarrow\frac{1}{3}\rightarrow\left|\leftarrow\frac{1}{3}\rightarrow\right|}{0 \qquad \frac{1}{3} \qquad \frac{2}{3} \qquad 1}$$

As before, 2 portions of hay are 2 parts of 3 parts, namely, $\frac{2}{3}$ of a bale, as well as two $\frac{1}{3}$'s, or $2(\frac{1}{3})$. Both expressions represent the same quantity, so that $2(\frac{1}{3}) = \frac{2}{3}$. In a like manner, $\frac{5}{6} = 5(\frac{1}{6})$ and $\frac{4}{7} = 4(\frac{1}{7})$, so that generally:

$$\frac{a}{b} = a\left(\frac{1}{b}\right) \quad \text{or} \quad \frac{a}{b} = \left(\frac{1}{b}\right) \cdot a \quad \text{where } a \text{ and } b \text{ are integers, } b \neq 0$$

The quantity $\frac{1}{b}$ is defined as the **multiplicative inverse** of b. A more common term is **reciprocal,** so that $\frac{1}{b}$ is the reciprocal of b.

This principle can now be stated as

or
    a over b is equal to a times the multiplicative inverse of b

    a over b is equal to a multiplied by the reciprocal of b

Before leaving these examples, note in (d) that $5(\frac{1}{5}) = 1$. Since $5(\frac{1}{5}) = \frac{5}{5}$, then $\frac{5}{5} = 1$. Similarly, in example (e), $3(\frac{1}{3}) = 1$ and $3(\frac{1}{3}) = \frac{3}{3}$, so that $\frac{3}{3} = 1$. In general, it is true that

$$a \cdot \frac{1}{a} = \frac{a}{a} = 1 \quad \text{and} \quad \frac{1}{a} \cdot a = \frac{a}{a} = 1 \quad \text{where } a \text{ is any integer, } a \neq 0$$

These relationships can be expressed by the statement:

    The product of any nonzero number with its multiplicative inverse is equal to one, the multiplicative identity element.

78   Pre-Algebra

Compare this with the statement in Chapter 2:

The sum of any number with its additive inverse is equal to zero, the additive identity element.

These two statements should impress the reader with their strong similarity.

Since the set of integers is a subset of the set of rational numbers, it is highly desirable that all properties of multiplication established for integers apply to multiplication for the set of rational numbers. Namely that:

The set of rational numbers is commutative with respect to multiplication.
The set of rational numbers is associative with respect to multiplication.

Henceforth, it shall be assumed that the set of rational numbers obeys these properties. Now that the groundwork has been established, rules for the multiplication of two rational numbers can be developed.

The product $\frac{1}{3} \cdot \frac{1}{2}$ can be established in several ways. First, by number line. Consider that the line segment from 0 to 1 is divided into 6 equal parts, 3 parts representing $\frac{1}{2}$ of the segment. Then, 1 of those 3 parts is $\frac{1}{3}$ of $\frac{1}{2}$, or 1 part of the 6 parts. Therefore, $\frac{1}{3} \cdot \frac{1}{2} = \frac{1}{6}$, as seen in the number line below:

(f) [number line from 0 to 1 divided into 6 equal parts of $\frac{1}{6}$, with $\frac{1}{2}$ marked]

Second,

$$6 \left(\frac{1}{3} \cdot \frac{1}{2}\right) = 3 \cdot 2 \left(\frac{1}{3} \cdot \frac{1}{2}\right)$$

$$= \left(3 \cdot \frac{1}{3}\right) \cdot \left(2 \cdot \frac{1}{2}\right) \quad \text{Using the commutative and associative properties}$$

$$= \left(\frac{3}{3}\right) \cdot \left(\frac{2}{2}\right)$$

$$= 1$$

Since $6 \cdot (\frac{1}{6}) = 1$, and $6(\frac{1}{3} \cdot \frac{1}{2}) = 1$, then $\frac{1}{3} \cdot \frac{1}{2} = \frac{1}{6}$.

Rational Numbers

Now consider a second product: $\frac{1}{4} \cdot \frac{1}{3}$. Using the number line, separate the line segment from 0 to 1 into 12 equal parts. Since $\frac{1}{3}$ is represented by 4 parts, and $\frac{1}{4}$ of 4 parts is 1, then $\frac{1}{4} \cdot \frac{1}{3}$ consists of 1 part in 12 parts, or: $\frac{1}{4} \cdot \frac{1}{3} = \frac{1}{12}$. This multiplication is illustrated in the diagram below. Observe that the word "of" is synonomous with multiplication.

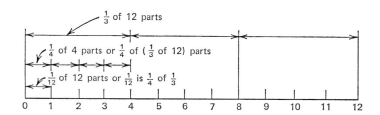

Again,

$$12 \cdot \left(\frac{1}{4} \cdot \frac{1}{3}\right) = 4 \cdot 3 \cdot \left(\frac{1}{4} \cdot \frac{1}{3}\right)$$

$$= \left(4 \cdot \frac{1}{4}\right) \cdot \left(3 \cdot \frac{1}{3}\right) \quad \text{Using the commutative and associative properties}$$

$$= \left(\frac{4}{4}\right) \cdot \left(\frac{3}{3}\right)$$

$$= 1$$

Since $12 \cdot \frac{1}{12} = 1$, and $12(\frac{1}{4} \cdot \frac{1}{3}) = 1$, then $\frac{1}{4} \cdot \frac{1}{3} = \frac{1}{12}$.

Observing that $\frac{1}{3} \cdot \frac{1}{2} = \frac{1}{6}$ and $\frac{1}{6} = \frac{1}{3 \cdot 2}$, then $\frac{1}{3} \cdot \frac{1}{2} = \frac{1}{3 \cdot 2}$. Also, $\frac{1}{4} \cdot \frac{1}{3} = \frac{1}{12}$ and $\frac{1}{12} = \frac{1}{4 \cdot 3}$, then $\frac{1}{4} \cdot \frac{1}{3} = \frac{1}{4 \cdot 3}$. So that generally:

$$\frac{1}{a} \cdot \frac{1}{b} = \frac{1}{a \cdot b} \quad \text{where } a \text{ and } b \text{ are integers, } a \neq 0 \text{ and } b \neq 0$$

Now consider the more general case with the product: $\frac{2}{3} \cdot \frac{4}{5}$. As in the previous example, the product can be found with the use of the number

80   Pre-Algebra

line, but it can be found more easily with the following arguments:

(g) $\dfrac{2}{3} \cdot \dfrac{4}{5} = \left(2 \cdot \dfrac{1}{3}\right) \cdot \left(4 \cdot \dfrac{1}{5}\right)$

$\phantom{\dfrac{2}{3} \cdot \dfrac{4}{5}} = 2 \cdot 4 \cdot \left(\dfrac{1}{3}\right)\left(\dfrac{1}{5}\right)$ Using the commutative and associative properties

$\phantom{\dfrac{2}{3} \cdot \dfrac{4}{5}} = 8 \cdot \left(\dfrac{1}{15}\right)$

$\dfrac{2}{3} \cdot \dfrac{4}{5} = \dfrac{8}{15}$

(h) $\dfrac{3}{7} \cdot \dfrac{5}{8} = \left(3 \cdot \dfrac{1}{7}\right)\left(5 \cdot \dfrac{1}{8}\right)$

$\phantom{\dfrac{3}{7} \cdot \dfrac{5}{8}} = 3 \cdot 5 \cdot \left(\dfrac{1}{7} \cdot \dfrac{1}{8}\right)$ Using the commutative and associative properties

$\phantom{\dfrac{3}{7} \cdot \dfrac{5}{8}} = 15 \cdot \left(\dfrac{1}{56}\right)$

$\dfrac{3}{7} \cdot \dfrac{5}{8} = \dfrac{15}{56}$

In example (g), $\dfrac{2}{3} \cdot \dfrac{4}{5} = \dfrac{8}{15}$, but also $\dfrac{8}{15} = \dfrac{2 \cdot 4}{3 \cdot 5}$, so that

$$\dfrac{2}{3} \cdot \dfrac{4}{5} = \dfrac{2 \cdot 4}{3 \cdot 5}$$

Again in example (h), $\dfrac{3}{7} \cdot \dfrac{5}{8} = \dfrac{15}{56}$, but also $\dfrac{15}{56} = \dfrac{3 \cdot 5}{7 \cdot 8}$, so that

$$\dfrac{3}{7} \cdot \dfrac{5}{8} = \dfrac{3 \cdot 5}{7 \cdot 8}$$

Many more examples can be furnished to convince the reader that generally:

$$\dfrac{a}{b} \cdot \dfrac{c}{d} = \dfrac{a \cdot c}{b \cdot d} \quad \text{for } a, b, c, \text{ and } d \text{ all integers, providing } b \neq 0, d \neq 0$$

Rational Numbers   81

Numbers of the form, $\frac{a}{b}$, where a and b are integers, $b \neq 0$, are defined as rational numbers, but they are also called **fractions.** When this term is used, "a" is referred to as the **numerator,** and "b" is referred to as the **denominator.** When fractions have a numerator of 1, they are called **unit fractions.** Using these terms:

> The product of two given rational numbers is equal to a rational number whose numerator is the product of the given numerators, and whose denominator is the product of the given denominators.

Note that the product $2(\frac{2}{3})$ can be found by treating it as a special case of the above rule.

(i) $\quad 2\left(\frac{2}{3}\right) = \frac{2}{1} \cdot \left(\frac{2}{3}\right) \qquad$ Since $\qquad 2 = \frac{2}{1}$

$\qquad\qquad = \frac{2 \cdot 2}{1 \cdot 3}$

$2 \cdot \left(\frac{2}{3}\right) = \frac{4}{3}$

Many more examples can be presented here, but perhaps it is now obvious that $4 \cdot (\frac{3}{7}) = \frac{12}{7}$, and $7 \cdot (\frac{5}{9}) = \frac{35}{9}$, so that generally:

$$a \cdot \left(\frac{b}{c}\right) = \frac{a \cdot b}{c} \qquad \text{where } a, b, \text{ and } c \text{ are integers, } c \neq 0$$

The reader may ask, what is the meaning of a negative rational number? To be consistent with the set of integers, it is desirable that the property, $-1 \cdot a = -a$, hold in the set of rational numbers. Thus, $-\frac{2}{3} = -1(\frac{2}{3})$, so that $-\frac{2}{3}$ is simply the opposite of 2 parts of the 3 parts of 1, or more specifically, 2 parts of the 3 parts of -1. This interpretation is shown in the diagram below:

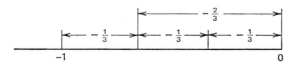

Pre-Algebra

Now consider these final examples:

(j) $\quad \dfrac{-2}{3} \cdot \dfrac{4}{7} = \dfrac{-2 \cdot 4}{3 \cdot 7}$

$\qquad\qquad = -\left(\left|\dfrac{-2}{3}\right| \cdot \left|\dfrac{4}{7}\right|\right)$

(k) $\quad \dfrac{5}{8} \cdot \dfrac{-3}{7} = \dfrac{5(-3)}{8 \cdot 7}$

$\qquad\qquad = -\left(\left|\dfrac{5}{8}\right| \cdot \left|\dfrac{-3}{7}\right|\right)$

(l) $\quad \dfrac{-7}{9} \cdot \dfrac{-3}{5} = \dfrac{(-7)(-3)}{9 \cdot 5}$

$\qquad\qquad = \left(\left|\dfrac{-7}{9}\right| \cdot \left|\dfrac{-3}{5}\right|\right)$

These three examples have illustrated the product of two rational numbers in which one is negative and the other is positive, or both are negative. The results are consistent with the laws of signs for the product of two integers given in Section 2.5, p. 63. It can be established that the following rules of signs apply for any two rational numbers:

1. The product of any two rational numbers with like signs is equal to the product of their absolute values.
2. The product of any two rational numbers with unlike signs is equal to the negative of the product of their absolute values.

Or:

1a. The product of any two rational numbers with like signs is a positive number.
2a. The product of any two rational numbers with unlike signs is a negative number.

3.1. Exercises

Perform the multiplication or simplification indicated for each of the following, writing your answer as an integer wherever possible:

1. $\dfrac{8}{1}$

2. $\dfrac{-5}{1}$

3. $4\left(\dfrac{1}{7}\right)$

4. $-3\left(\dfrac{1}{8}\right)$

5. $7\left(\dfrac{2}{5}\right)$

6. $-13\left(\dfrac{3}{4}\right)$

7. $29 \cdot \left(\dfrac{1}{29}\right)$

8. $-637 \cdot \left(\dfrac{1}{-637}\right)$

9. $3\left(\dfrac{2}{5}\right)$

10. $-4\left(\dfrac{3}{7}\right)$

11. $7 \cdot \left(\dfrac{2}{7}\right)$

Hint: $7\left(\dfrac{2}{7}\right) = 7\left(\dfrac{1}{7} \cdot 2\right)$

12. $-5\left(\dfrac{+2}{-5}\right)$

13. $47\left(\dfrac{-3}{47}\right)$

14. $637\left(\dfrac{37}{637}\right)$

15. $\dfrac{2}{3} \cdot \dfrac{4}{7}$

16. $\dfrac{3}{5} \cdot \dfrac{2}{11}$

17. $\dfrac{5}{7} \cdot \dfrac{8}{9}$

18. $\dfrac{7}{3} \cdot \dfrac{8}{5}$

19. $\dfrac{-4}{9} \cdot \dfrac{2}{3}$

20. $\dfrac{8}{9} \cdot \dfrac{-5}{7}$

21. $\dfrac{-3}{8} \cdot \dfrac{-5}{7}$

22. $\dfrac{-6}{5} \cdot \dfrac{-3}{5}$

23. $\dfrac{-8}{-9} \cdot \dfrac{-4}{-7}$

24. $\left(-\dfrac{7}{16}\right) \cdot \left(-\dfrac{3}{2}\right)$

25. $-\left(\dfrac{4}{7} \cdot \dfrac{8}{9}\right)$

26. $\dfrac{1}{5} \cdot \left(\dfrac{3}{-4}\right)$

27. $\left(-\dfrac{2}{3}\right) \cdot \left(-\dfrac{5}{9}\right)$

28. $3 \cdot \left(\dfrac{-2}{5}\right) \cdot 7\left(\dfrac{4}{5}\right)$

29. $-7\left(\dfrac{4}{-7}\right) \cdot 8\left(\dfrac{-3}{8}\right)$

30. $83\left(\dfrac{-7}{83}\right) \cdot 217\left(\dfrac{3}{217}\right)$

31. $-5\left(\dfrac{3}{7}\right) 3 \cdot \left(\dfrac{-5}{7}\right)$

32. $-2 \cdot \left(\dfrac{5}{7}\right)\left[-2 \cdot \left(\dfrac{-4}{9}\right)\right]$

84   Pre-Algebra

Write the multiplicative inverse of each of the following integers:

33. $+4$ 34. -3 35. $6-9$ 36. $12-(-3)$

Find the reciprocal of each of the following:

37. -8 38. 5 39. $\dfrac{1}{2}$ 40. $-\dfrac{1}{7}$

3.2. Fundamental Property of Fractions

The concept of the reciprocal of a number was introduced in the previous section, but only the reciprocals of integers and unit fractions were found there. The following discussion will lead to a simple method of finding the reciprocal of any nonzero rational number.

It was previously determined that the product of any number with its reciprocal is 1, the multiplicative identity element. Restating this in symbolic terms, $a \cdot \dfrac{1}{a} = 1$, as developed in Section 3.1, p. 77. Now consider the statement: $\tfrac{2}{3} \cdot \tfrac{3}{2} = 1$. Since the product of these two rational numbers is 1, $\tfrac{2}{3}$ and $\tfrac{3}{2}$ are reciprocals. From above, the reciprocal of any number "a" is also $\dfrac{1}{a}$, so that the reciprocal of $\tfrac{2}{3}$ is also $\dfrac{1}{\tfrac{2}{3}}$. It can be proved that each nonzero number can only have one reciprocal, so that $\tfrac{3}{2} = \dfrac{1}{\tfrac{2}{3}}$. Similarly, $\tfrac{2}{3} = \dfrac{1}{\tfrac{3}{2}}$. Again, $\tfrac{5}{7} \cdot \tfrac{7}{5} = 1$, so that $\tfrac{5}{7}$ and $\tfrac{7}{5}$ are reciprocals, but also $\tfrac{5}{7} = \dfrac{1}{\tfrac{7}{5}}$ and $\tfrac{7}{5} = \dfrac{1}{\tfrac{5}{7}}$.

Restating the above:

If $\dfrac{a}{b}$ is any rational number, $a \neq 0$ and $b \neq 0$, then its multiplicative inverse, or reciprocal, is $\dfrac{b}{a}$, so that $\dfrac{a}{b} \cdot \dfrac{b}{a} = 1$, and also, $\dfrac{a}{b} = \dfrac{1}{\tfrac{b}{a}}$ as well as $\dfrac{b}{a} = \dfrac{1}{\tfrac{a}{b}}$.

Rational Numbers 85

Using this definition:

The reciprocal of $\frac{7}{8}$ is $\frac{8}{7}$.
The reciprocal of $-\frac{4}{5}$ is $-\frac{5}{4}$.
The reciprocal of 3, or $\frac{3}{1}$, is $\frac{1}{3}$.

Since no rational number can be found whose product with 0 is 1, 0 has no reciprocal and $\frac{1}{0}$ has no meaning. Likewise, $\frac{2}{0}$, $\frac{3}{0}$, and in general, $\frac{a}{0}$, $a \neq 0$, has no meaning and is said to be **undefined**.

Rational numbers may have a negative numerator, a negative denominator, or the entire number may be negative. Consider the following fractions:

(a) $-\frac{2}{3}$ (b) $\frac{-4}{7}$ (c) $\frac{5}{-8}$ (d) $-\frac{-1}{6}$

Referring to (a), observe the following argument:

$$-\frac{2}{3} = -1 \cdot \frac{2}{3}$$ $\qquad -a = -1 \cdot a$

$$= -1 \cdot \left(2 \cdot \frac{1}{3}\right)$$ $\qquad \frac{a}{b} = a \cdot \frac{1}{b}$

$$= (-1 \cdot 2) \cdot \frac{1}{3}$$ $\qquad$ Associative property of multiplication

$$= -2 \cdot \frac{1}{3}$$

$$-\frac{2}{3} = \frac{-2}{3}$$ $\qquad a \cdot \frac{1}{b} = \frac{a}{b}$

Using this same result for example (b), $\frac{-4}{7} = -\frac{4}{7}$.

Examples of the relationship, $a \cdot \frac{1}{a} = 1$, stated in Section 3.1, p. 77, were furnished for positive integers, and only assumed to be true for negative integers. To demonstrate that the relationship holds for negative integers, observe that $\frac{-1}{1}$ and $\frac{1}{-1}$ are reciprocals, so that $\frac{-1}{1} \cdot \frac{1}{-1} = 1$,

Pre-Algebra

and $\dfrac{-1}{-1} = 1$, by the multiplication of two rational numbers. This can be generalized to confirm that $a \cdot \dfrac{1}{a} = \dfrac{a}{a} = 1$, for any nonzero integer.

Now referring to example (c),

$$\dfrac{5}{-8} = \dfrac{5}{-8} \cdot 1 \qquad\qquad a \cdot 1 = a$$

$$= \dfrac{5}{-8} \cdot \dfrac{-1}{-1} \qquad\qquad \dfrac{-1}{-1} = 1$$

$$\dfrac{5}{-8} = \dfrac{-5}{8} \qquad\qquad \text{Multiplication of two rational numbers}$$

Example (d) can be simplified as follows:

$$-\dfrac{-1}{6} = -\left(-\dfrac{1}{6}\right) \qquad\qquad \text{Since } \dfrac{-1}{6} = -\dfrac{1}{6}$$

$$= \dfrac{1}{6} \qquad\qquad -(-a) = a$$

More examples could be presented to establish the relationship that

$$\dfrac{-a}{b} = \dfrac{a}{-b} = -\dfrac{a}{b} \qquad \text{where } a \text{ and } b \text{ are any two integers, } b \neq 0$$

This statement illustrates the principle that the value of a fraction remains the same if:

1. The signs of both the numerator and denominator are changed to their opposites.

Or:

2. The sign of either the numerator or denominator and the sign of the fraction are changed to their opposites.

Some examples will help to make this statement clear:

(e) $\quad -\dfrac{-4}{7} = \dfrac{-(-4)}{7} = \dfrac{4}{7}$

(f) $\dfrac{3}{-8} = -\dfrac{3}{8} = \dfrac{-3}{8}$

(g) $-\dfrac{-5}{-9} = \dfrac{-5}{-(-9)} = \dfrac{-5}{9}$

Although mathematicians offer no rigid rule, it is generally agreed that a fraction should be expressed with no more than one negative sign. Whenever it is necessary to use a negative sign, it is generally written in the numerator.

Now, some additional examples illustrating an important concept:

(h) $\dfrac{6}{8} = \dfrac{3 \cdot 2}{4 \cdot 2}$

$\phantom{\dfrac{6}{8}} = \dfrac{3}{4} \cdot \dfrac{2}{2}$ $\qquad \dfrac{a \cdot c}{b \cdot d} = \dfrac{a}{b} \cdot \dfrac{c}{d}$

$\phantom{\dfrac{6}{8}} = \dfrac{3}{4} \cdot 1$ $\qquad$ Since $\dfrac{2}{2} = 1$

$\phantom{\dfrac{6}{8}} = \dfrac{3}{4}$ $\qquad a \cdot 1 = a$

(i) $\dfrac{-16}{32} = \dfrac{-1 \cdot 16}{2 \cdot 16}$

$\phantom{\dfrac{-16}{32}} = \dfrac{-1}{2} \cdot \dfrac{16}{16}$ $\qquad \dfrac{a \cdot c}{b \cdot d} = \dfrac{a}{b} \cdot \dfrac{c}{d}$

$\phantom{\dfrac{-16}{32}} = \dfrac{-1}{2} \cdot 1$ $\qquad$ Since $\dfrac{16}{16} = 1$

$\phantom{\dfrac{-16}{32}} = \dfrac{-1}{2}$ $\qquad a \cdot 1 = a$

(j) $\dfrac{5}{9} = \dfrac{5}{9} \cdot 1$ $\qquad a = a \cdot 1$

$\phantom{\dfrac{5}{9}} = \dfrac{5}{9} \cdot \dfrac{3}{3}$ $\qquad \dfrac{3}{3} = 1$

$\phantom{\dfrac{5}{9}} = \dfrac{15}{27}$ $\qquad \dfrac{a}{b} \cdot \dfrac{c}{d} = \dfrac{a \cdot c}{b \cdot d}$

88   Pre-Algebra

(k) $\dfrac{7}{-3} = \dfrac{7}{-3} \cdot 1$ $a = a \cdot 1$

$\phantom{(k)\ \dfrac{7}{-3}} = \dfrac{7}{-3} \cdot \dfrac{-2}{-2}$ Since $1 = \dfrac{-2}{-2}$

$\phantom{(k)\ \dfrac{7}{-3}} = \dfrac{-14}{6}$ $\dfrac{a}{b} \cdot \dfrac{c}{d} = \dfrac{a \cdot c}{b \cdot d}$

Observe in (h) and (i) that the greatest common factor, or GCF, was factored from each of these numbers. The resulting fraction is simplified, in that the absolute values of the numerator and denominator are smaller than in the original fraction. Whenever all common factors are removed from a fraction, the fraction is said to be **reduced.** In (j) and (k), the opposite result was achieved, in that the absolute values of the numerator and denominator were made larger. These results can be generalized and stated as:

(i) $\dfrac{a}{b} = \dfrac{a \cdot c}{b \cdot c}$

(ii) $\dfrac{a \cdot c}{b \cdot c} = \dfrac{a}{b}$ where a, b, and c are integers, $b \neq 0$, $c \neq 0$

This principle is called the **fundamental property of fractions,** and its importance will become obvious later in this chapter.

If the numerator and denominator of a fraction are referred to as its two parts, then by using (i) of the fundamental property of fractions, one part of the fraction can be multiplied by a factor if the second part is also multiplied by the same factor. Using (ii) of this property, a factor can be removed from any part of a fraction, as long as the same factor is removed from the remaining part. Two fractions which differ only by common factors in the numerator and denominator are said to be **equivalent fractions.** For example, $\tfrac{1}{2}$ and $\tfrac{3}{6}$ are equivalent fractions because $\tfrac{1}{2} = \dfrac{1 \cdot 3}{2 \cdot 3}$.

Thus, equivalent fractions can either be formed or identified by use of the fundamental property of fractions.

Rational Numbers   89

Consider the following examples to illustrate a final relationship before concluding this section:

(1) $8\left(\dfrac{7}{8}\right) = 8 \cdot \left(7 \cdot \dfrac{1}{8}\right)$ $\dfrac{a}{b} = a \cdot \dfrac{1}{b}$

$= 7 \cdot \left(8 \cdot \dfrac{1}{8}\right)$ Commutative and associative properties

$= 7 \cdot 1$ $a \cdot \dfrac{1}{a} = 1$

$= 7$ $a \cdot 1 = a$

(m) $-3\left(\dfrac{2}{-3}\right) = -3\left(2 \cdot \dfrac{1}{-3}\right)$ $\dfrac{a}{b} = a \cdot \dfrac{1}{b}$

$= 2 \cdot \left(-3 \cdot \dfrac{1}{-3}\right)$ Commutative and associative properties

$= 2 \cdot 1$ $a \cdot \dfrac{1}{a} = 1$

$= 2$ $a \cdot 1 = a$

These last two examples should serve to make it clear that

$b \cdot \dfrac{a}{b} = a$ and $\dfrac{a}{b} \cdot b = a$ for a and b any two integers, $b \neq 0$

Using this property, the following multiplications can be performed mentally:

$7 \cdot \dfrac{4}{7} = 4$ $-6 \cdot \left(\dfrac{5}{-6}\right) = 5$ $4 \cdot \left(\dfrac{-3}{4}\right) = -3$

$2{,}785 \cdot \left(\dfrac{1{,}783}{2{,}785}\right) = 1{,}783$

3.2. Exercises

Find the reciprocals for each of the following rational numbers:

1. $\dfrac{4}{7}$

2. $\dfrac{-9}{5}$

3. $-\dfrac{1}{3}$

4. -6

5. $\dfrac{1}{\frac{7}{8}}$

6. $\dfrac{1}{\frac{-1}{4}}$

Pre-Algebra

Simplify each of the following fractions, leaving at most one negative sign:

7. $-\dfrac{-2}{3}$
8. $\dfrac{-4}{-7}$
9. $-\dfrac{5}{-9}$
10. $-\dfrac{-6}{-5}$

Reduce each of the following fractions by removing the GCF by the fundamental property of fractions:

11. $\dfrac{-9}{6}$
12. $\dfrac{14}{56}$
13. $\dfrac{-14}{4}$
14. $\dfrac{15}{35}$
15. $\dfrac{-27}{-49}$
16. $\dfrac{78}{104}$
17. $\dfrac{-33}{88}$
18. $\dfrac{68}{-51}$
19. $\dfrac{21}{-49}$
20. $\dfrac{-105}{63}$
21. $\dfrac{-36}{-84}$
22. $\dfrac{105}{28}$
23. $\dfrac{-48}{168}$
24. $\dfrac{285}{76}$
25. $\dfrac{-120}{-125}$
26. $\dfrac{87}{171}$

Supply the missing numerator or denominator to make each of the following pairs of rational numbers equal:

27. $\dfrac{2}{3} = \dfrac{}{18}$
28. $\dfrac{4}{7} = \dfrac{}{-49}$
29. $\dfrac{8}{-3} = \dfrac{24}{}$
30. $\dfrac{-5}{6} = \dfrac{30}{}$
31. $\dfrac{8}{15} = \dfrac{-24}{}$
32. $\dfrac{-7}{17} = \dfrac{21}{}$
33. $\dfrac{5}{7} = \dfrac{}{-49}$
34. $\dfrac{5}{-14} = \dfrac{}{42}$
35. $\dfrac{42}{8} = \dfrac{}{4}$
36. $\dfrac{17}{3} = \dfrac{}{-27}$
37. $\dfrac{-63}{-56} = \dfrac{}{8}$
38. $\dfrac{147}{-91} = \dfrac{-21}{}$
39. $\dfrac{-16}{20} = \dfrac{4}{}$
40. $\dfrac{15}{40} = \dfrac{}{-8}$
41. $\dfrac{28}{42} = \dfrac{2}{}$
42. $\dfrac{78}{90} = \dfrac{-13}{}$
43. $\dfrac{57}{76} = \dfrac{3}{}$
44. $\dfrac{4}{7} = \dfrac{-92}{}$

Rational Numbers 91

Perform each of the following multiplications **mentally**:

45. $8 \cdot \dfrac{3}{8}$

46. $-4 \cdot \left(\dfrac{3}{-4}\right)$

47. $23 \cdot \left(\dfrac{-17}{23}\right)$

48. $2{,}637{,}015 \cdot \left(\dfrac{78{,}156}{2{,}637{,}015}\right)$

49. $-6\left(\dfrac{-5}{6}\right)$

50. $8 \cdot \left(\dfrac{3}{-8}\right)$

3.3. Common Factors in Multiplication

Before continuing to other operations with rational numbers, it would be well to discuss an additional aid to multiplication. Mathematicians generally reduce rational numbers to aid in calculations. For example, it is far easier to deal with $\frac{1}{2}$ rather than $\frac{276}{552}$. The multiplication $\frac{161}{483} \cdot \frac{37}{259}$ can be performed with greater ease when all common factors are removed, and the multiplication then becomes $\frac{1}{3} \cdot \frac{1}{7}$. In general, it is far easier to perform multiplications by removing all common factors first. A few examples should make this point clear:

(a) First method, multiplying first:

$$14 \cdot \frac{2}{7} = \frac{28}{7}$$
$$= \frac{4 \cdot 7}{1 \cdot 7}$$
$$= \frac{4}{1}$$
$$= 4$$

$$a \cdot \frac{b}{c} = \frac{a \cdot b}{c}$$

$$\frac{a \cdot c}{b \cdot c} = \frac{a}{b}$$

Second method, removing common factors first:

$$14 \cdot \frac{2}{7} = 2\left(7 \cdot \frac{2}{7}\right) \qquad \text{Associative property of multiplication}$$
$$= 2 \cdot 2$$
$$= 4$$

$$a \cdot \frac{b}{a} = b$$

Pre-Algebra

Third method, striking out common factors first:

$$14 \cdot \frac{2}{7} = \overset{2}{\cancel{14}} \cdot \frac{2}{\underset{1}{\cancel{7}}} \qquad \text{Striking out 7 in the numerator and denominator}$$

$$= 2 \cdot \frac{2}{1}$$

$$= 4$$

When the second method is understood, much time can be saved by striking out the common factors as indicated in the third method.

(b) First method, multiplying first:

$$42 \cdot \frac{3}{-8} = \frac{126}{-8} \qquad\qquad a \cdot \frac{b}{c} = \frac{a \cdot b}{c}$$

$$= \frac{63 \cdot 2}{-4 \cdot 2}$$

$$= \frac{63}{-4} \qquad\qquad \text{Fundamental property of fractions}$$

$$= \frac{-63}{4} \qquad\qquad \frac{a}{-b} = \frac{-a}{b}$$

Second method, removing common factors first:

$$42 \cdot \frac{3}{-8} = 21 \cdot 2 \cdot \left(\frac{1}{2} \cdot \frac{3}{-4}\right)$$

$$= 21 \cdot \frac{3}{-4} \cdot \left(2 \cdot \frac{1}{2}\right) \qquad \text{Commutative and associative properties}$$

$$= \frac{63}{-4} \cdot 1 \qquad\qquad a \cdot \frac{b}{c} = \frac{a \cdot b}{c} \quad \text{and} \quad a \cdot \frac{1}{a} = 1$$

$$= \frac{-63}{4} \qquad\qquad a \cdot 1 = a \quad \text{and} \quad \frac{a}{-b} = \frac{-a}{b}$$

Rational Numbers   93

Third method, striking out the common factors first:

$$42 \cdot \frac{3}{-8} = {}^{21}\cancel{42} \cdot \frac{3}{\underset{-4}{\cancel{-8}}}$$ Striking out a 2 in the numerator and denominator

$$= \frac{63}{-4}$$ $$a \cdot \frac{b}{c} = \frac{a \cdot b}{c}$$

$$= \frac{-63}{4}$$ $$\frac{a}{-b} = \frac{-a}{b}$$

This method can also be utilized when multiplying two fractions, as indicated in the following examples:

(c) $$\frac{3}{8} \cdot \frac{4}{7} = \left(\frac{3}{2} \cdot \frac{1}{4}\right) \cdot \left(4 \cdot \frac{1}{7}\right)$$

$$= \frac{3}{2} \cdot \frac{1}{7} \cdot \left(\frac{1}{4} \cdot 4\right)$$

$$= \frac{3}{14} \cdot 1$$ $$\frac{1}{a} \cdot a = 1$$

$$= \frac{3}{14}$$

Again, striking out the common factors first:

$$\frac{3}{8} \cdot \frac{4}{7} = \frac{3}{{}_2\cancel{8}} \cdot \frac{\cancel{4}^1}{7}$$

$$= \frac{3 \cdot 1}{2 \cdot 7}$$

$$= \frac{3}{14}$$

(d) $$\frac{5}{-14} \cdot \frac{49}{15} = \left(\frac{1}{-2} \cdot \frac{5}{7}\right)\left(\frac{7}{5} \cdot \frac{7}{3}\right)$$

$$= \left(\frac{1}{-2} \cdot \frac{7}{3}\right)\left(\frac{5}{7} \cdot \frac{7}{5}\right)$$

$$= \frac{7}{-6} \cdot 1$$

$$= \frac{-7}{6}$$

Pre-Algebra

Again, striking out the common factors first:

$$\frac{5}{-14} \cdot \frac{49}{15} = \frac{\overset{1}{\cancel{5}}}{_{-2}\cancel{-14}} \cdot \frac{\overset{7}{\cancel{49}}}{\cancel{15}_{3}}$$

$$= \frac{1 \cdot 7}{-2 \cdot 3}$$

$$= \frac{7}{-6}$$

$$= \frac{-7}{6}$$

The method of striking out common factors before multiplication is probably familiar to most students under the name of "canceling." Recently, teachers of mathematics have avoided this term in the belief that it suggests an improper use of the concept. Henceforth, it will not be used again in this text in the hope that the use of the term "common factors" will strongly suggest to the student to strike out a common factor in the numerator and denominator. Some additional examples using this technique follow:

(e) $\dfrac{9}{26} \cdot \dfrac{4}{42} = \dfrac{9}{26} \cdot \dfrac{\overset{2}{\cancel{4}}}{\cancel{42}_{21}}$

$\phantom{(e)\ \dfrac{9}{26} \cdot \dfrac{4}{42}} = \dfrac{\overset{3}{\cancel{9}}}{13\cancel{26}} \cdot \dfrac{\overset{1}{\cancel{2}}}{\cancel{21}_{7}}$

$\phantom{(e)\ \dfrac{9}{26} \cdot \dfrac{4}{42}} = \dfrac{3}{91}$

(f) $\dfrac{8}{34} \cdot \dfrac{9}{28} = \dfrac{\overset{4}{\cancel{8}}}{17\cancel{34}} \cdot \dfrac{9}{28}$

$\phantom{(f)\ \dfrac{8}{34} \cdot \dfrac{9}{28}} = \dfrac{\overset{1}{\cancel{4}}}{17} \cdot \dfrac{9}{\cancel{28}_{7}}$

$\phantom{(f)\ \dfrac{8}{34} \cdot \dfrac{9}{28}} = \dfrac{1 \cdot 9}{17 \cdot 7}$

$\phantom{(f)\ \dfrac{8}{34} \cdot \dfrac{9}{28}} = \dfrac{9}{119}$

Rational Numbers   95

One final word of caution: When using the technique of striking out common factors when multiplying, be careful to strike out one common factor in the numerator with one common factor in the denominator. Common factors can be in the numerator and denominator of one fraction, or in the numerator of one fraction and denominator of a second, as in examples (e) and (f). Factors in two numerators or two denominators are **not** considered to be common factors.

3.3. Exercises

Reduce each of the following fractions:

1. $\dfrac{12}{28}$
2. $\dfrac{70}{105}$
3. $\dfrac{36}{210}$
4. $\dfrac{-42}{104}$
5. $\dfrac{-330}{396}$
6. $\dfrac{105}{-36}$
7. $\dfrac{168}{-16}$
8. $\dfrac{316}{-231}$

Multiply each of the following by first removing all common factors. There should be no common factors in the product:

9. $42 \cdot \dfrac{3}{8}$
10. $-24 \cdot \dfrac{5}{9}$
11. $43 \cdot \left(\dfrac{-6}{21}\right)$
12. $13 \cdot \left(\dfrac{35}{-14}\right)$
13. $22 \cdot \dfrac{6}{132}$
14. $42\left(\dfrac{-81}{105}\right)$
15. $-45\left(\dfrac{154}{-70}\right)$
16. $-36\left(\dfrac{-33}{-42}\right)$

17. $\dfrac{12}{15} \cdot \dfrac{3}{8}$
18. $\dfrac{5}{9} \cdot \dfrac{54}{15}$
19. $\dfrac{3}{4} \cdot \dfrac{28}{35}$
20. $\dfrac{25}{36} \cdot \dfrac{48}{45}$
21. $\dfrac{49}{64} \cdot \dfrac{12}{14}$
22. $\dfrac{27}{63} \cdot \dfrac{35}{33}$
23. $\dfrac{45}{77} \cdot \dfrac{63}{25}$
24. $\dfrac{36}{75} \cdot \dfrac{105}{99}$

25. $\dfrac{98}{28} \cdot \dfrac{80}{70}$
26. $\dfrac{42}{30} \cdot \dfrac{35}{49}$
27. $\dfrac{120}{25} \cdot \dfrac{135}{150}$
28. $\dfrac{375}{430} \cdot \dfrac{550}{500}$
29. $\left(\dfrac{24}{40} \cdot \dfrac{28}{27}\right) \cdot \dfrac{15}{49}$
30. $\dfrac{56}{35} \cdot \left(\dfrac{65}{72} \cdot \dfrac{85}{39}\right)$
31. $\dfrac{231}{88} \cdot \dfrac{54}{28} \cdot \dfrac{112}{69}$
32. $\dfrac{74}{38} \cdot \dfrac{95}{231} \cdot \dfrac{147}{87} \cdot \dfrac{58}{111}$

3.4. Division of Rational Numbers

Quite frequently, it is desired to separate certain quantities into a given number of parts. Consider the following examples:

(a) Mr. Smith passed away and left $12,000 to be apportioned so that each of his four grandchildren receives an equal amount of money. How much money does each grandchild get?

(b) Irene drove 495 miles on a recent trip and used 33 gallons of gasoline. How many miles did she travel per gallon of gasoline used?

The process of separating quantities into a given number of equal parts is called the binary operation of **division,** which is indicated by the symbol "$\div$" and called a division sign. If 1 unit of length is marked off into two equal parts, then 6 of these parts is 3 units of length. This can be stated as $\frac{6}{2} = 3$. It is also true that 6 units of length can be divided into 2 equal parts so that each part is 3 units of length. This is stated as $6 \div 2 = 3$. Since both methods yield 3, $\frac{6}{2} = 6 \div 2$. The two diagrams below should help to illustrate that these two processes do produce the same result:

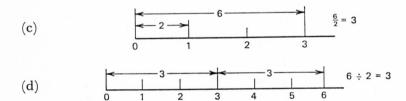

(c)

(d)

Therefore, in general, $\frac{a}{b}$ and $a \div b$ are equivalent expressions. Although the previous examples were limited to natural numbers, it can be demonstrated that

$$\frac{a}{b} = a \div b \quad \text{where } a \text{ and } b \text{ are any two rational numbers, } b \neq 0$$

The result of dividing two numbers is called the **quotient.** The quotient of a and b, or $a \div b$, is read "a divided by b," with b called the **divisor** and a the **dividend.** Because of the above property, any division of rational numbers can be treated as a fraction, or vice versa. The following examples

Rational Numbers 97

will illustrate this equality:

(e) $\dfrac{2}{3} \div 4 = \dfrac{\frac{2}{3}}{4}$ (f) $5 \div \dfrac{3}{4} = \dfrac{5}{\frac{3}{4}}$ (g) $\dfrac{4}{7} \div \dfrac{3}{8} = \dfrac{\frac{4}{7}}{\frac{3}{8}}$

In (e), $\frac{2}{3}$ is the numerator of the resulting fraction, and 4 is the denominator. In (f), 5 is the numerator and $\frac{3}{4}$ is the denominator. In (g), $\frac{4}{7}$ is the numerator and $\frac{3}{8}$ is the denominator. Note in each case that a longer line is used to separate the numerator and the denominator of the fraction. This is called the **main fraction bar.** The following examples will serve to develop a simple way to divide any two rational numbers:

(h) $2 \div 3 = \dfrac{2}{3}$ $a \div b = \dfrac{a}{b}$

$ 2 \div 3 = 2 \cdot \dfrac{1}{3}$ $\dfrac{a}{b} = a \cdot \dfrac{1}{b}$

(i) $4 \div \dfrac{3}{7} = \dfrac{4}{\frac{3}{7}}$ $a \div b = \dfrac{a}{b}$

$ = 4 \cdot \dfrac{1}{\frac{3}{7}}$ $\dfrac{a}{b} = a \cdot \dfrac{1}{b}$

$ 4 \div \dfrac{3}{7} = 4 \cdot \dfrac{7}{3}$ $\dfrac{1}{\frac{a}{b}} = \dfrac{b}{a}$

(j) $\dfrac{5}{8} \div \dfrac{4}{3} = \dfrac{\frac{5}{8}}{\frac{4}{3}}$ $a \div b = \dfrac{a}{b}$

$ = \dfrac{5}{8} \cdot \dfrac{1}{\frac{4}{3}}$ $\dfrac{a}{b} = a \cdot \dfrac{1}{b}$

$ \dfrac{5}{8} \div \dfrac{4}{3} = \dfrac{5}{8} \cdot \dfrac{3}{4}$ $\dfrac{1}{\frac{a}{b}} = \dfrac{b}{a}$

Observe in (h) that $2 \div 3$ is equal to 2 times the reciprocal of 3, or $\frac{1}{3}$. In (i), $4 \div \frac{3}{7}$ becomes 4 times the reciprocal of $\frac{3}{7}$, or $\frac{7}{3}$. Again, in (j), $\frac{5}{8} \div \frac{4}{3}$ is equal to $\frac{5}{8}$ times the reciprocal of $\frac{4}{3}$ or $\frac{3}{4}$. It should be clear that each of these division problems can be replaced by an equivalent statement involving multiplication. This can be stated by the principle

$$a \div b = \frac{a}{b} = a \cdot \frac{1}{b} \qquad \text{for any two rational numbers, } b \neq 0$$

or

$\qquad$ *a* divided by *b* is equal to *a* times the reciprocal of *b*

or once more:

$\qquad$ The quotient of two numbers is equal to the product of the first and the multiplicative inverse of the second.

Now compare this to the definition of subtraction given in Section 2.3, p. 52, namely:

$\qquad$ The difference of two numbers is equal to the sum of the first and the additive inverse of the second.

Hopefully, these parallel statements will impress the reader that the relationship between multiplication and division is almost the same as that between addition and subtraction. Many more fascinating relationships such as this exist in mathematics.

The following examples can now be easily simplified using this new principle:

(k) $\quad 7 \div \frac{5}{7} = 7 \cdot \frac{7}{5} = \frac{49}{5}$

(l) $\quad \frac{3}{4} \div 5 = \frac{3}{4} \cdot \frac{1}{5} = \frac{3}{20}$

(m) $\quad \frac{2}{9} \div \frac{5}{7} = \frac{2}{9} \cdot \frac{7}{5} = \frac{14}{45}$

Observe that:

(n) $\quad 1 \cdot 7 \div 7 = 1 \qquad$ and $\qquad 1 \div 7 \cdot 7 = 1$

(o) $\quad 1 \cdot \frac{2}{3} \div \frac{2}{3} = 1 \qquad$ and $\qquad 1 \div \frac{2}{3} \cdot \frac{2}{3} = 1$

Rational Numbers

These examples illustrate that the original number remains after the two operations, multiplication and division, are performed in either order. That is, one operation "undoes" the second so that multiplication and division are inverse operations, as previously defined in Section 2.4, p. 58.

Mixed operations involving division can be confusing and require some discussion. Consider these examples:

(p) $7 \div 9 \cdot 3$

    First method: $(7 \div 9) \cdot 3 = \dfrac{7}{3}$

    Second method: $7 \div (9 \cdot 3) = 7 \div 27 = \dfrac{7}{27}$

(q) $8 \div 4 + 2$

    First method: $(8 \div 4) + 2 = 2 + 2 = 4$

    Second method: $8 \div (4 + 2) = 8 \div 6 = \dfrac{8}{6} = \dfrac{4}{3}$

(r) $9 - 6 \div 3$

    First method: $(9 - 6) \div 3 = 3 \div 3 = 1$

    Second method: $9 - (6 \div 3) = 9 - \dfrac{6}{3} = 9 - 2 = 7$

(s) $9 \div 3 + 2 \cdot 8 - 4 = \dfrac{9}{3} + 16 - 4$

$\qquad\qquad\qquad\qquad\;\; = 3 + 12$

$\qquad\qquad\qquad\qquad\;\; = 15$

Examples (p), (q), and (r) all demonstrate that results will vary depending upon the order in which the operations are performed. Again, some arbitrary decision must be made to avoid conflicting answers. In expressions involving any combination of the four operations, mathematicians agree that:

> Multiplications and/or divisions are always performed before additions and/or subtractions. Unless grouping symbols indicate otherwise, multiplications and divisions are performed from left to right. This rule also holds for additions and subtractions.

In (p) and (q) then, the first method is correct. The second method is correct in (r). Example (s) is simplified correctly in that the division and multiplication are performed first, then the addition and subtraction.

Some final comments about division:

Since $7 \div 9 \neq 9 \div 7$ and $2 \div \frac{3}{4} \neq \frac{3}{4} \div 2$, division of nonequal rational numbers is not a commutative operation.

Since $(7 \div 9) \div 3 = \frac{7}{9} \div 3 = \frac{7}{9} \cdot \frac{1}{3} = \frac{2}{27}$ and

$$7 \div (9 \div 3) = 7 \div 3 = \frac{7}{3}$$

therefore,

$$(7 \div 9) \div 3 \neq 7 \div (9 \div 3)$$

so that division is not an associative operation.

Since $7 \div 1 = 7$ but $1 \div 7 \neq 7$, 1 is not the identity element for division and it can be shown that no identity element exists for division.

Since $3 \div 2$ is not an integer, as well as countless other quotients of two integers, the set of integers is not closed with respect to division. The extension of the integers to the set of rational numbers provides a set which is now closed with respect to addition, subtraction, multiplication, and division. A new operation shall provide the motivation to extend the rational numbers once again in the next chapter.

3.4. Exercises

Perform the indicated divisions. Express the quotients as rational numbers with all the common factors removed.

1. $18 \div 4$

2. $-27 \div (-18)$

3. $\frac{2}{5} \div 9$

4. $\frac{-3}{4} \div 7$

5. $\frac{-4}{7} \div 8$

6. $\frac{5}{9} \div (-25)$

Rational Numbers 101

7. $\dfrac{3}{4} \div \dfrac{8}{7}$

8. $\dfrac{-5}{9} \div \dfrac{-8}{5}$

9. $\dfrac{4}{5} \div \dfrac{20}{17}$

10. $\dfrac{-3}{7} \div \dfrac{12}{21}$

11. $\dfrac{28}{17} \div \dfrac{-21}{85}$

12. $\dfrac{-12}{75} \div \dfrac{33}{-35}$

13. $\dfrac{14}{15} \div \dfrac{49}{70}$

14. $\dfrac{42}{81} \div \dfrac{45}{108}$

15. $\dfrac{33}{65} \div \dfrac{63}{91}$

16. $\dfrac{-40}{35} \div \dfrac{110}{125}$

Simplify each of the following expressions involving mixed operations. Express your result as a rational number which is reduced:

17. $7 \div 4 \div 3$
18. $(-4) - 9 \cdot 2$
19. $-9 \div 3 + 2$
20. $8 - 5 \cdot 6 \div 2$
21. $5 - 27 \div 3$
22. $17 - 12 \div 2 \cdot 3$
23. $8 \cdot 7 \div 4 - 15 \cdot 3 \div 9$
24. $(9 - 21) \div (23 - 8)$
25. $9 \div (6 \div 7)$
26. $(9 \div 6) \div 7$
27. $-3 - 14 \div (-2)$
28. $[15 \div (-5) - 7] \div (-2)$
29. $20 \div 5 - 6 \cdot 7$
30. $17 - 12 \div 4 - 5 \cdot 7$
31. $\dfrac{9}{7} \div (5 \div 3)$
32. $\dfrac{8}{5} \cdot (9 - 3) \div \dfrac{7}{5} \cdot (6 - 8)$
33. $-28 \div [-12 \div (-3) + 3]$
34. $[35 - 7 \div (-3) \cdot 6] \div (-7)$
35. $(43 - 7) \div (5 - 9) \div (1 - 19)$
36. $9 \cdot 8 \div 6 - 5 \cdot 7$
37. $\dfrac{4}{9} \div 8 \div \dfrac{7}{3}$
38. $14 - 14 \div \dfrac{7}{6} - \dfrac{3}{8} \div \dfrac{9}{48}$
39. $6 - 9 \cdot 7 \div 3$
40. $42 \div \dfrac{14}{5} \div \dfrac{45}{7} \div \dfrac{35}{8}$

3.5. The Division Rule

Before continuing to the remaining operations with rational numbers, a few final comments regarding division of rational numbers are in order.

Since the division of rational numbers in Section 3.4 was restated as multiplication with the inverse element, no new rules were developed for the sign of the quotient of two rational numbers. The rule for the signs of the product of two rational numbers was used instead. Rules do exist for the division of two rational numbers, and since they are occasionally useful, they will be developed here.

Since $6 \cdot 2 = 12$, it also holds that $12 \div 2 = 6$. That is, the product of two numbers divided by one of its factors is equal to the remaining factor. These results can be generalized to form the statement:

> If $a \cdot b = p$, then $a = p \div b$, and $b = p \div a$, where a, b, and p are nonzero rational numbers.

By using this statement:

(a) If $-4 \cdot (-3) = 12$, then $12 \div (-3) = -4$.
(b) If $-3 \cdot (4) = -12$, then $-12 \div 4 = -3$.
(c) If $4(-3) = -12$, then $-12 \div (-3) = 4$.

In (a) and (b), the quotient of a positive and a negative number is a negative number. In (c), the quotient of two negative numbers is a positive number. The quotient of two positive numbers is obvious, so that:

> If two rational numbers have like signs, their quotient is positive.
> If two rational numbers have unlike signs, their quotient is negative.

Note the replacing "quotient" with "product," these rules are identical to those of the multiplication of two rational numbers. They will be referred to hereafter as the **division rule.**

The following examples will illustrate this rule:

(d) $-8 \div (-2) = +4$ The division rule of like signs
(e) $+9 \div (-3) = -3$ The division rule of unlike signs
(f) $-10 \div 2 = -5$ The division rule of unlike signs

Rational Numbers 103

Continuing, note the following set of statements:

(g) $\quad 0 \cdot 6 = 0,\quad$ so that $\quad 0 \div 6 = 0$
$\quad\quad 0 \cdot (-3) = 0,\quad$ so that $\quad 0 \div (-3) = 0$

$$0 \cdot \frac{1}{4} = 0, \quad \text{so that} \quad 0 \div \frac{1}{4} = 0$$

These statements illustrate the fact that 0 divided by any nonzero rational number is 0. Stated symbolically:

$$\frac{0}{a} = 0 \quad \text{where } a \text{ is any rational number, } a \neq 0$$

Since $\frac{a}{0}$, $a \neq 0$, is undefined, as discussed in Section 3.2, p. 85:

Division by 0 is undefined.

Since $6 \cdot 0 = 0$ and $8 \cdot 0 = 0$, it follows that $6 = \frac{0}{0}$ and $8 = \frac{0}{0}$. It therefore follows, strangely enough, that $\frac{0}{0}$ can be any number and for this reason, it is called an **indeterminate** expression. In summation then, $\frac{a}{0}$, $a \neq 0$, has no equivalent, but $\frac{a}{0}$, $a = 0$, has too many equivalents.

If the division of two natural numbers does not yield a natural number, the quotient may be expressed as the sum of a natural number and a fraction. The following examples will illustrate this:

(h) $\quad \dfrac{19}{3} = 19 \cdot \dfrac{1}{3}$

$\quad\quad = (18 + 1) \cdot \dfrac{1}{3}$

$\quad\quad = 18 \cdot \dfrac{1}{3} + 1 \cdot \dfrac{1}{3} \quad\quad\quad$ By the distributive property

$\quad\quad = 6 + \dfrac{1}{3}$

(i) $\dfrac{263}{27} = 263 \cdot \dfrac{1}{27}$

$= (243 + 20) \cdot \dfrac{1}{27}$

$= 243 \cdot \dfrac{1}{27} + 20 \cdot \dfrac{1}{27}$ By the distributive property

$= 9 + \dfrac{20}{27}$

(j) $\dfrac{3{,}564}{47} = 3{,}564 \cdot \dfrac{1}{47}$

$= (3{,}525 + 39) \cdot \dfrac{1}{47}$

$= 3{,}525 \cdot \dfrac{1}{47} + 39 \cdot \dfrac{1}{47}$ By the distributive property

$= 75 + \dfrac{39}{47}$

The results, $6 + \tfrac{1}{3}$, $9 + \tfrac{20}{27}$, and $75 + \tfrac{39}{47}$, above, can be written as $6\tfrac{1}{3}$, $9\tfrac{20}{27}$, and $75\tfrac{39}{47}$, respectively. These expressions are called **mixed numbers.** The process above yields the correct solution, but it is not the easiest method available. **Long division,** the algorithm currently in use, is the favored method, and it should be familiar enough to most students not to require an extensive explanation. Examples (h), (i), and (j) can now be demonstrated using the long division algorithm, which is restricted to natural numbers:

(h)
```
      6
   ─────
 3)19
   18
   ──
    1
```

(i)
```
      9
    ─────
 27)263
    242
    ───
     20
```

(j)
```
       75
    ──────
 47)3,564
    3 29
    ────
     274
     235
     ───
      39
```

The numbers 1, 20, and 39, which remain, are called the **remainders,** and in each case they must be less than the divisor. The numbers 6, 9, and

75, are called the **partial quotients**. In order to express the complete quotient, the partial quotient is added to the remainder, which is written over the divisor as a fraction.

In general, the quotient of a and b can be written as

$$\frac{a}{b} = q + \frac{r}{b} \qquad \text{where } a \text{ and } b \text{ are natural numbers, } b \neq 0 \text{ and}$$
$$0 \leq r < b$$

In the expression above, q is the partial quotient, $\frac{r}{b}$ is the fractional part of the quotient, and r is the remainder. When $r = 0$, a is said to be **exactly divisible** by b.

This author feels that calculations in algebra are much more difficult with mixed numbers rather than their equivalent fractions. When algebraic symbols are written next to each other, they indicate multiplication. The mixed number expression indicates that the two parts are to be added, thus creating confusion when algebraic symbols are present. Unless physical measures such as distance and weight are involved, it is strongly urged that mixed numbers not be used here or in the following mathematics courses.

3.5. Exercises

Simplify each of the following quotients by using the division rule. Express your answer as a reduced rational number:

1. $16 \div (-4)$
2. $-18 \div (-3)$
3. $-42 \div 14$
4. $-[63 \div (-9)]$
5. $-[(-28) \div 7]$
6. $-[-56 \div (-21)]$
7. $-56 \div (-36)$
8. $-[124 \div (84)]$
9. $-\left(\frac{10}{7} \div 4\right)$
10. $\frac{35}{-8} \div 63$

106   Pre-Algebra

11. $-9 \div \left(\dfrac{-6}{5}\right)$

12. $42 \div \dfrac{14}{81}$

13. $\dfrac{-4}{13} \div (-12)$

14. $\dfrac{35}{-27} \div 49$

15. $36 \div \left(\dfrac{-4}{7}\right)$

16. $-56 \div \left(\dfrac{-32}{10}\right)$

17. $\dfrac{18}{45} \div \dfrac{54}{35}$

18. $\left(-\dfrac{4}{9}\right) \div \left(-\dfrac{12}{42}\right)$

19. $\left(\dfrac{-26}{55} \div 4\right) \div \dfrac{7}{44}$

20. $\dfrac{-15}{63} \div \left[\left(-\dfrac{8}{5}\right) \div \dfrac{42}{25}\right]$

21. $\dfrac{-28}{63} \div \dfrac{21}{-27}$

22. $-\left[\dfrac{216}{147} \div \dfrac{360}{157}\right]$

23. $-\left[\dfrac{-198}{180} \div \dfrac{121}{75}\right]$

24. $\dfrac{-81}{21} \div (39) \div \dfrac{27}{52}$

Simplify each of the following quotients by long division. If the quotient is a mixed number, remove all common factors from the fractional part:

25. $432 \div 6$
26. $578 \div 8$
27. $4{,}048 \div 23$
28. $7{,}317 \div (+39)$
29. $12{,}385 \div (314)$
30. $85{,}753 \div (1{,}305)$

3.6. Least Common Multiple

Before discussing the addition and subtraction of rational numbers, it will first be necessary to introduce the concept of the **least common multiple,** which is abbreviated as the **LCM.**

The **multiple** of a given number is simply the given number multiplied by any integer. For example, 6 is a multiple of 3, 9 is a multiple of 3, and generally, the set $\{0,3,6,9,12,\ldots\}$ consists of all the multiples of 3 greater than or equal to 0. Likewise:

(a) $E = \{0,2,4,6,8,10,\ldots\}$ represents the multiples of 2 greater than or equal to 0

(b) $F = \{0,5,10,15,20,\ldots\}$ represents the multiples of 5 greater than or equal to 0

Now observe that 6 is a multiple of 3 and also a multiple of 2, so that 6 is said to be a **common multiple** of 2 and 3. In addition to 6, the numbers, 12, 18, 24, etc., are also common multiples of 2 and 3. Since 6 is the smallest of all the common multiples, it is referred to as the **least common multiple,** or LCM of 2 and 3. Specifically then:

> The LCM of A, any finite set of natural numbers, is the smallest natural number which is a multiple of all elements in A.

To further illustrate this concept, 15 is a multiple of 3, and 15 is a multiple of 5. In addition, 30, 45, 60, etc., are common multiples, but 15 is the LCM. The common multiples of 8 and 24 are {24,48,72,96,. . .} and the LCM is 24.

It shall soon be obvious how and where the LCM is used, but first it is desirable to develop a simple method of finding this number. In order to do so, consider the factors of 8, which are $1 \cdot 8$ and $2 \cdot 4$. Neither of these factorizations are complete, in that, both can be factored again to arrive at the product: $2 \cdot 2 \cdot 2$. When a natural number greater than 1 has no factors other than 1 and itself, it is called a **prime number.** Examples of prime numbers are: 2, 3, 5, 7, 11, 13, 17, etc. Note that 1 is not considered to be a prime number, so that 2 is the smallest prime number. Natural numbers which are not prime are called **composite** numbers. Examples of composite numbers are: 4, 6, 8, 9, 10, 12, 14, 15, 18, 20, 21, etc. When each of the factors of a number is a prime number, then the number is said to be **completely factored.** Thus, examples of numbers which are completely factored are:

(c) $90 = 9 \cdot 10 = 3 \cdot 3 \cdot 2 \cdot 5 = 2 \cdot 3 \cdot 3 \cdot 5$
(d) $315 = 5 \cdot 63 = 5 \cdot 7 \cdot 9 = 3 \cdot 3 \cdot 5 \cdot 7$

When two numbers have no factors in common, they are said to be **relatively prime.** Examples of pairs of numbers which are relatively prime are: 6, 7 and 4, 9. Observe that neither number need be prime in order for the pair of numbers to be relatively prime. When a fraction is completely reduced, the numerator and denominator are relatively prime. That is, they have no factors in common. It can also be said that two numbers are relatively prime when their greatest common factor is 1.

Now to find the LCM of two numbers using the concept of the complete factorization of these numbers.

108   Pre-Algebra

To find the LCM of 12 and 18, first completely factor each of these numbers. Therefore, $12 = 2 \cdot 6 = 2 \cdot 2 \cdot 3$ and $18 = 2 \cdot 9 = 2 \cdot 3 \cdot 3$. Now find the union of these sets of factors, representing each factor the greatest number of times that it occurs in any one number. The LCM of the two numbers is the product of this new set of factors. Therefore, the LCM of 12 and 18 is $2 \cdot 2 \cdot 3 \cdot 3$, or 36. Note that 36 is a multiple of 12 and a multiple of 18, and the least common multiple of these numbers. Some additional examples illustrating this idea follow:

(e)   Find the LCM of 8 and 12:

$8 = 2 \cdot 2 \cdot 2$ The LCM is $2 \cdot 2 \cdot 2 \cdot 3$, which is
$12 = 2 \cdot 2 \cdot 3$ 24

(f)   Find the LCM of 90 and 189:

$90 = 2 \cdot 3 \cdot 3 \cdot 5$ The LCM is $2 \cdot 3 \cdot 3 \cdot 3 \cdot 5 \cdot 7$,
$189 = 9 \cdot 21 = 3 \cdot 3 \cdot 3 \cdot 7$ which is 1890

The least common multiple of two denominators is called the **least common denominator,** which is abbreviated as the **LCD.** The following example demonstrates the mechanics of finding the LCD, however its usefulness shall not become apparent until the next section:

(g)   Find the LCD of $\frac{1}{21}$ and $\frac{1}{18}$:

$21 = 3 \cdot 7$ The LCM of 21 and 18 is $2 \cdot 3 \cdot 3 \cdot 7 = 126$
$18 = 2 \cdot 3 \cdot 3$ The LCD of $\frac{1}{21}$ and $\frac{1}{18}$ is 126

This LCD is the smallest multiple of both the given denominators.

3.6.   Exercises

Find the complete factorization for each of the following numbers:

1. 24
2. 60
3. 98
4. 232
5. 630
6. 4,950
7. 1,001
8. 2,310
9. 945
10. 2,640
11. 7,605
12. 714

Rational Numbers 109

Find the LCM for each of the following sets of numbers:

13. {9,63}
14. {10,21}
15. {6,9}
16. {9,21}
17. {15,35}
18. {28,105}
19. {20,330}
20. {210,378}
21. {90,675}
22. {315,735}
23. {24,15,108}
24. {105,45,175}

Find the least common denominator (LCD) for each of the following sets of fractions:

25. $\left\{\dfrac{1}{8}, \dfrac{1}{12}\right\}$

26. $\left\{\dfrac{1}{10}, \dfrac{1}{35}\right\}$

27. $\left\{\dfrac{1}{12}, \dfrac{1}{18}\right\}$

28. $\left\{\dfrac{1}{18}, \dfrac{1}{27}\right\}$

29. $\left\{\dfrac{1}{26}, \dfrac{1}{75}\right\}$

30. $\left\{\dfrac{5}{48}, \dfrac{3}{56}\right\}$

31. $\left\{\dfrac{7}{39}, \dfrac{13}{75}\right\}$

32. $\left\{\dfrac{17}{45}, \dfrac{8}{63}\right\}$

33. $\left\{\dfrac{1}{28}, \dfrac{7}{441}\right\}$

34. $\left\{\dfrac{2}{315}, \dfrac{7}{3375}\right\}$

35. $\left\{\dfrac{7}{98}, \dfrac{3}{35}, \dfrac{3}{40}\right\}$

36. $\left\{\dfrac{2}{75}, \dfrac{7}{45}, \dfrac{1}{54}\right\}$

37. Find all the prime numbers less than 50.
38. Find all the prime numbers greater than 50 and less than 100.

3.7. Addition and Subtraction of Rational Numbers

There are many instances when it is necessary to find the sum or difference of two rational numbers. For example:

(a) If Sherry ate $\frac{1}{6}$ of a pie and Gary ate $\frac{1}{4}$ of the same pie, what part of the pie did they consume?

(b) If Norm ran $\frac{1}{5}$ of a mile and walked $\frac{3}{4}$ of a mile, what is the total distance that he traveled?

(c) A man left $\frac{1}{2}$ of his inheritance to his wife, and $\frac{3}{8}$ to his son. What part of his estate was given to the remainder of his relatives?

These and many other examples demonstrate the need for a method of adding or subtracting any two rational numbers. Addition will be developed in two cases: adding two rationals with like denominators, and adding two rationals with unlike denominators. The following examples will illustrate the first case:

(d) $\dfrac{1}{7} + \dfrac{3}{7} = 1 \cdot \dfrac{1}{7} + 3 \cdot \dfrac{1}{7}$ $\dfrac{a}{b} = a \cdot \dfrac{1}{b}$

$\phantom{\dfrac{1}{7} + \dfrac{3}{7}} = (1 + 3)\dfrac{1}{7}$ Distributive property

$\phantom{\dfrac{1}{7} + \dfrac{3}{7}} = \dfrac{1 + 3}{7}$ $a \cdot \dfrac{1}{b} = \dfrac{a}{b}$

$\phantom{\dfrac{1}{7} + \dfrac{3}{7}} = \dfrac{4}{7}$

(e) $\dfrac{3}{9} + \dfrac{2}{9} = 3 \cdot \dfrac{1}{9} + 2 \cdot \dfrac{1}{9}$ $\dfrac{a}{b} = a \cdot \dfrac{1}{b}$

$\phantom{\dfrac{3}{9} + \dfrac{2}{9}} = (3 + 2) \cdot \dfrac{1}{9}$ Distributive property

$\phantom{\dfrac{3}{9} + \dfrac{2}{9}} = \dfrac{3 + 2}{9}$ $a \cdot \dfrac{1}{b} = \dfrac{a}{b}$

$\phantom{\dfrac{3}{9} + \dfrac{2}{9}} = \dfrac{5}{9}$

Note that $\dfrac{1}{7} + \dfrac{3}{7} = \dfrac{1+3}{7}$ and $\dfrac{3}{9} + \dfrac{2}{9} = \dfrac{3+2}{9}$, so that in general,

$$\dfrac{a}{c} + \dfrac{b}{c} = \dfrac{a+b}{c} \quad \text{for } a, b, \text{ and } c \text{ any three rational numbers, } c \neq 0$$

Restating this principle:

> The sum of two rational numbers with like denominators is a rational number whose denominator is the same as the given denominator, and the numerator is the sum of the numerators of the two given rational numbers.

Rational Numbers 111

Now consider the following, which are examples of the second case:

(f) $\dfrac{1}{3} + \dfrac{1}{2} = \dfrac{1 \cdot 2}{3 \cdot 2} + \dfrac{1 \cdot 3}{2 \cdot 3}$ The LCM of 2 and 3 is 6

$= \dfrac{2}{6} + \dfrac{3}{6}$

$= \dfrac{2 + 3}{6}$ $\dfrac{a}{c} + \dfrac{b}{c} = \dfrac{a + b}{c}$

$= \dfrac{5}{6}$

(g) $\dfrac{5}{6} + \dfrac{3}{8} = \dfrac{5 \cdot 4}{6 \cdot 4} + \dfrac{3 \cdot 3}{8 \cdot 3}$ The LCM of 6 and 8 is 24

$= \dfrac{20}{24} + \dfrac{9}{24}$

$= \dfrac{20 + 9}{24}$ $\dfrac{a}{c} + \dfrac{b}{c} = \dfrac{a + b}{c}$

$= \dfrac{29}{24}$

The last two examples illustrate the principle that:

> When adding two rational numbers with unlike denominators, first find the LCM of the two denominators, and then change each rational number to an equivalent fraction whose denominator is the LCM. Now add the two fractions with like denominators as above.

Several more examples will help to make this clear:

(h) $\dfrac{5}{9} + \dfrac{7}{24} = \dfrac{5 \cdot 8}{9 \cdot 8} + \dfrac{7 \cdot 3}{24 \cdot 3}$ The LCM of 9 and 24 is 72

$= \dfrac{40}{72} + \dfrac{21}{72}$

$= \dfrac{61}{72}$ $\dfrac{a}{c} + \dfrac{b}{c} = \dfrac{a + b}{c}$

Pre-Algebra

(i) $\dfrac{9}{24} + \dfrac{17}{40} = \dfrac{45}{120} + \dfrac{51}{120}$

$= \dfrac{96}{120}$

$= \dfrac{4}{5}$

The subtraction of rational numbers is developed much the same as above, but using the subtraction principle, $a - b = a + (-b)$, to first change the given problem into an addition problem. Subtraction is developed in two cases as in addition. Consider the following two examples illustrating the case in which the denominators are alike:

(j) $\dfrac{7}{9} - \dfrac{3}{9} = \dfrac{7}{9} + \left(-\dfrac{3}{9}\right)$ $\qquad a - b = a + (-b)$

$= \dfrac{7}{9} + \left(\dfrac{-3}{9}\right)$ $\qquad -\dfrac{a}{b} = \dfrac{-a}{b}$

$= \dfrac{7 + (-3)}{9}$ $\qquad \dfrac{a}{c} + \dfrac{b}{c} = \dfrac{a+b}{c}$

$= \dfrac{7 - 3}{9}$

$= \dfrac{4}{9}$

(k) $\dfrac{-4}{15} - \dfrac{8}{15} = \dfrac{-4}{15} + \left(-\dfrac{8}{15}\right)$ $\qquad a - b = a + (-b)$

$= \dfrac{-4}{15} + \left(\dfrac{-8}{15}\right)$ $\qquad -\dfrac{a}{b} = \dfrac{-a}{b}$

$= \dfrac{-4 + (-8)}{15}$

$= \dfrac{-4 - 8}{15}$

$= \dfrac{-12}{15}$

$= \dfrac{-4}{5}$

Rational Numbers

Note that $\dfrac{7}{9} - \dfrac{3}{9} = \dfrac{7-3}{9}$ and $\dfrac{-4}{15} - \dfrac{8}{15} = \dfrac{-4-8}{15}$, so the general principle can be stated as

$$\dfrac{a}{c} - \dfrac{b}{c} = \dfrac{a-b}{c}$$ where a, b, and c are any three rational numbers, $c \neq 0$

Without using symbolism, the statement becomes:

The difference of two rational numbers with like denominators is a rational number having the same denominator, and a numerator which is the difference of the numerators of the two given rational numbers.

Actually, the statement $\dfrac{a}{c} + \dfrac{b}{c} = \dfrac{a+b}{c}$ applies to both addition (where b is positive) and subtraction (where b is negative), since $\dfrac{a-b}{c} = \dfrac{a+(-b)}{c}$.

The following examples illustrate the case in which both denominators are unlike:

(1) $\dfrac{6}{7} - \dfrac{4}{5} = \dfrac{6}{7} + \left(-\dfrac{4}{5}\right)$

$= \dfrac{6}{7} + \left(\dfrac{-4}{5}\right)$ $\qquad -\dfrac{a}{b} = \dfrac{-a}{b}$

$= \dfrac{30}{35} + \left(\dfrac{-28}{35}\right)$ Changing 7 and 5 to a LCM of 35

$= \dfrac{30 + (-28)}{35}$

$= \dfrac{2}{35}$

(m) $\dfrac{-5}{6} - \dfrac{1}{9} = \dfrac{-5}{6} + \left(-\dfrac{1}{9}\right)$

$= \dfrac{-15}{18} + \dfrac{-2}{18}$ Changing 6 and 9 to a LCM of 18

$= \dfrac{-15 + (-2)}{18}$

$= \dfrac{-17}{18}$

By first changing the difference of two rational numbers to an addition problem, the addition can then be performed with the addition rule of two rational numbers with unlike denominators.

If an integer is added to or subtracted from a rational number, first change the integer to $\dfrac{a}{1}$, then change the two denominators to the LCM, and add. For example:

(n) $3 + \dfrac{7}{5} = \dfrac{3}{1} + \dfrac{7}{5}$

$= \dfrac{3 \cdot 5}{1 \cdot 5} + \dfrac{7}{5}$ 5 is the LCM of 5 and 1

$= \dfrac{15 + 7}{5}$

$= \dfrac{22}{5}$

(o) $-4 + \dfrac{3}{4} = \dfrac{-4}{1} + \dfrac{3}{4}$

$= \dfrac{-16}{4} + \dfrac{3}{4}$ 4 is the LCM of 4 and 1

$= \dfrac{-16 + 3}{4}$

$= \dfrac{-13}{4}$

The convention for mixed operations given in Section 3.4, p. 99, applies to all rational numbers. For example:

(p) $\frac{2}{3} \cdot \frac{4}{3} + \frac{2}{27} \div \frac{5}{9} = \frac{8}{9} + \frac{2}{27} \cdot \frac{9}{5}$ Multiply and divide first

$= \frac{8}{9} + \frac{2}{15}$

$= \frac{8 \cdot 5}{9 \cdot 5} + \frac{2 \cdot 3}{15 \cdot 3}$ The LCM of 9 and 15 is 45

$= \frac{40 + 6}{45}$

$= \frac{46}{45}$

Since $\frac{1}{7} + \frac{2}{7} = \frac{3}{7}$ and $\frac{2}{7} + \frac{1}{7} = \frac{3}{7}$, then $\frac{1}{7} + \frac{2}{7} = \frac{2}{7} + \frac{1}{7}$. Also,

$$\left(\frac{1}{7} + \frac{2}{7}\right) + \frac{3}{7} = \frac{6}{7} \quad \text{and} \quad \frac{1}{7} + \left(\frac{2}{7} + \frac{3}{7}\right) = \frac{6}{7}$$

so that

$$\left(\frac{1}{7} + \frac{2}{7}\right) + \frac{3}{7} = \frac{1}{7} + \left(\frac{2}{7} + \frac{3}{7}\right)$$

The principles in the last two examples can be generalized to the statement:

> The addition of two rational numbers is a commutative operation, and the addition of three rational numbers is an associative operation.

The identity element for the addition of rational numbers can be established to be 0, and it can also be established that subtraction of rational numbers has no identity element.

The following example demonstrates that the distributive property also holds for rational numbers:

(q) First adding:

$$\frac{2}{3}\left(\frac{5}{8} + \frac{1}{6}\right) = \frac{2}{3}\left(\frac{15}{24} + \frac{4}{24}\right)$$

$$= \frac{2}{3} \cdot \frac{19}{24}$$

$$= \frac{19}{36}$$

116   Pre-Algebra

First multiplying:
$$\frac{2}{3}\left(\frac{5}{8} + \frac{1}{6}\right) = \frac{2}{3} \cdot \frac{5}{8} + \frac{2}{3} \cdot \frac{1}{6}$$
$$= \frac{5}{12} + \frac{1}{9}$$
$$= \frac{15}{36} + \frac{4}{36}$$
$$= \frac{19}{36}$$

Or generally:
$$a \cdot (b + c) = a \cdot b + a \cdot c \quad \text{where } a, b, \text{ and } c \text{ are any three rational numbers}$$

3.7. Exercises

Add or subtract each of the following as indicated. Reduce your answers:

1. $\frac{5}{7} + \frac{2}{7}$

2. $\frac{3}{16} + \frac{9}{16}$

3. $\frac{7}{15} + \frac{2}{15}$

4. $\frac{1}{21} + \frac{13}{21}$

5. $\frac{9}{8} - \frac{3}{8}$

6. $\frac{7}{12} - \frac{5}{12}$

7. $\frac{5}{18} - \frac{17}{18}$

8. $\frac{5}{9} - \frac{20}{9}$

9. $\frac{3}{8} + \frac{1}{4}$

10. $\frac{5}{3} + \frac{2}{9}$

11. $\frac{3}{2} - \frac{5}{6}$

12. $\frac{3}{5} - \frac{17}{20}$

13. $\frac{2}{7} + \frac{3}{4}$

14. $\frac{3}{8} + \frac{5}{12}$

15. $\frac{7}{15} + \frac{23}{35}$

16. $\frac{8}{45} + \frac{7}{36}$

17. $\frac{9}{16} - \frac{5}{12}$

18. $\frac{-3}{20} - \frac{7}{32}$

19. $\frac{8}{15} - \frac{11}{20}$

20. $\frac{-6}{35} + \frac{7}{20}$

21. $\frac{3}{28} + \frac{5}{36}$

22. $\frac{7}{18} - \frac{4}{75}$

23. $\frac{18}{57} - \frac{42}{78}$
(Hint: reduce first)

24. $\frac{9}{69} + \frac{8}{115}$

25. $\frac{-7}{45} + \frac{5}{-54}$

26. $\frac{8}{51} - \frac{-2}{9}$

27. $\frac{-3}{44} - \frac{5}{32}$

28. $\frac{8}{58} - \frac{14}{34}$

29. $6 + \frac{3}{8}$

30. $3 + \frac{5}{6}$

31. $7 - \frac{4}{7}$

32. $4 - \frac{5}{9}$

33. $-4 + \frac{5}{6}$

Rational Numbers 117

34. $-6 + \dfrac{3}{4}$ 35. $-3 - \dfrac{2}{5}$ 36. $-5 - \dfrac{2}{9}$

Simplify each of the following mixed operations:

37. $\dfrac{6}{14} + \dfrac{5}{21} - \dfrac{2}{35}$

38. $\dfrac{8}{15} \div \dfrac{7}{5} - \dfrac{2}{35}$

39. $\dfrac{5}{9} - \dfrac{1}{6} + \dfrac{3}{5} \cdot \dfrac{7}{9}$

40. $\dfrac{2}{3} \cdot \dfrac{4}{7} + \dfrac{5}{8} \div \dfrac{3}{2}$

41. $4 \div 7 - 3 \div 2 + 5 \div 9$

42. $6 \div (4 - 9) \div (8 - 11)(9 - 6) - (3 - 11) \div (8 - 3)$

43. $8 \div \left(\dfrac{3}{4}\right) \div \dfrac{7}{3} \div \left(\dfrac{4}{18} \cdot \dfrac{3}{24}\right) \div \dfrac{16}{3}$

44. $\dfrac{7}{6} - 8 \cdot 4 \div 5 \cdot 3 + 5 \div 21$

45. $3 \div 7 - 7 \div 3$
46. $12 - 7 \div 5 - 3(-4) \div 15$
47. $3 + 4(-6) \div 7 - 5(3) \div (8 - 16)$

48. $28 \div \left(\dfrac{7}{-3}\right) \div \dfrac{4}{-5} \div (6 - 14) \div (-2 - 5)$

3.8. Order of Rational Numbers

There are many occasions when it is desired to compare two rational numbers to determine which is the largest. In this section, a simple method is proposed to find which of two given rational numbers is the largest. In Section 1.7, p. 34, it was determined that if b is to the right of a on the number line, then $b > a$. The same principle holds for two rational numbers. That is, since $\frac{4}{7}$ represents 4 parts of 7 and $\frac{3}{7}$ represents 3 parts of 7, $\frac{4}{7}$ is to the right of $\frac{3}{7}$ and therefore, $\frac{4}{7} > \frac{3}{7}$.

Refer to the diagram below:

(a)

By the same reasoning, $\frac{5}{9} > \frac{3}{9}$, $\frac{7}{13} > \frac{6}{13}$, etc., and in general:

If $a > b$, then $\frac{a}{c} > \frac{b}{c}$, provided that $c > 0$.

When comparing two rational numbers whose denominators are unlike, first change the given fractions to equivalent fractions having the LCD, then compare the numerators as before. For example:

(b) Compare $\frac{3}{4}$ and $\frac{4}{5}$:

$$\frac{3}{4} = \frac{15}{20} \quad \text{and} \quad \frac{4}{5} = \frac{16}{20} \qquad \text{The LCM of 4 and 5 is 20}$$

So that

$$\frac{15}{20} < \frac{16}{20} \qquad \text{Since } 15 < 16$$

and therefore,

$$\frac{3}{4} < \frac{4}{5}$$

(c) Compare $\frac{-4}{7}$ and $\frac{-3}{8}$:

$$\frac{-4}{7} = \frac{-32}{56} \quad \text{and} \quad \frac{-3}{8} = \frac{-21}{56} \qquad \text{The LCM of 7 and 8 is 56}$$

Therefore

$$\frac{-32}{56} < \frac{-21}{56} \qquad \text{Since } -32 < -21$$

and

$$\frac{-4}{7} < \frac{-3}{8}$$

It is a remarkable fact that between any two rational numbers, no matter how close together they are on the number line, there can be found at least one, and actually infinitely many rational numbers. This is shown in the following examples:

(d) Find a rational number between $\frac{3}{7}$ and $\frac{4}{7}$.

Since $\frac{3}{7} = \frac{6}{14}$ and $\frac{4}{7} = \frac{8}{14}$, the task is to find a number between $\frac{6}{14}$ and $\frac{8}{14}$. A rational number satisfying this requirement is $\frac{7}{14}$ or $\frac{1}{2}$, so that $\frac{3}{7} < \frac{1}{2} < \frac{4}{7}$.

(e) Find a rational number between $\frac{14}{9}$ and $\frac{15}{8}$.

Since $\frac{14}{9} = \frac{112}{72}$ and $\frac{15}{8} = \frac{135}{72}$, $\frac{120}{72} = \frac{5}{3}$ is a number which will satisfy the given condition. Therefore, $\frac{112}{72} < \frac{120}{72} < \frac{135}{72}$ and so $\frac{14}{9} < \frac{5}{3} < \frac{15}{8}$.

3.8. Exercises

Find the correct relationship ($>$, $=$, $<$) between each of the following pairs of numbers:

1. $\frac{5}{6}$ ___ $\frac{2}{3}$
2. $\frac{5}{8}$ ___ $\frac{3}{4}$
3. $\frac{9}{10}$ ___ $\frac{900}{10,000}$
4. $\frac{3}{7}$ ___ $\frac{5}{14}$
5. $\frac{28}{18}$ ___ $\frac{42}{27}$
6. $\frac{7}{6}$ ___ $\frac{9}{8}$
7. $\frac{1}{2}$ ___ $\frac{1}{3}$
8. $\frac{3}{5}$ ___ $\frac{3}{4}$
9. $\frac{4}{7}$ ___ $\frac{5}{8}$
10. $\frac{7}{5}$ ___ $\frac{5}{7}$
11. $\frac{3}{13}$ ___ $\frac{24}{104}$
12. $\frac{16}{9}$ ___ $\frac{21}{12}$
13. $\frac{5}{24}$ ___ $\frac{1}{5}$
14. $\frac{11}{18}$ ___ $\frac{7}{11}$
15. $\frac{17}{5}$ ___ $\frac{7}{2}$
16. $\frac{5}{9}$ ___ $\frac{4}{7}$
17. $\frac{6}{28}$ ___ $\frac{7}{35}$
18. $\frac{8}{3}$ ___ $\frac{5}{2}$
19. $\frac{6}{11}$ ___ $\frac{3}{5}$
20. $\frac{9}{16}$ ___ $\frac{5}{9}$

Pre-Algebra

Find a rational number for each of the following pairs which is greater than the first, and less than the second:

21. $\dfrac{3}{4}, \dfrac{7}{8}$ 23. $\dfrac{-3}{5}, \dfrac{-5}{9}$ 25. $\dfrac{-4}{7}, \dfrac{-1}{2}$ 27. $\dfrac{8}{19}, \dfrac{3}{7}$

22. $\dfrac{9}{4}, \dfrac{5}{2}$ 24. $\dfrac{8}{13}, \dfrac{5}{8}$ 26. $\dfrac{7}{12}, \dfrac{3}{5}$ 28. $\dfrac{-4}{9}, \dfrac{-7}{16}$

Find three rational numbers for each of the following pairs which are larger than the first, and less than the second:

29. $\dfrac{1}{2}, \dfrac{3}{4}$ 30. $\dfrac{1}{7}, \dfrac{1}{4}$ 31. $\dfrac{7}{12}, \dfrac{5}{8}$ 32. $\dfrac{-5}{6}, \dfrac{-4}{5}$

Order the following numbers with the smallest first, and the largest last:

33. $\dfrac{4}{7}, \dfrac{1}{3}, \dfrac{1}{2}, -\dfrac{1}{6}, \dfrac{8}{21}$ 35. $\dfrac{3}{5}, \dfrac{5}{7}, \dfrac{2}{3}, \dfrac{7}{11}$

34. $\dfrac{1}{2}, \dfrac{5}{12}, \dfrac{1}{4}, \dfrac{1}{6}, \dfrac{1}{3}, \dfrac{1}{12}$ 36. $\dfrac{17}{21}, \dfrac{5}{6}, \dfrac{11}{14}, \dfrac{8}{9}, \dfrac{7}{8}$

3.9. Compound Fractions

A rational number has been defined as any number of the form $\dfrac{a}{b}$, where a is any integer and b is a nonzero integer. If the numerator and denominator of a rational number are integers, the rational number is usually called a **simple fraction.** When the numerator and/or the denominator of a rational number are simple fractions, then the expression is referred to as a **compound fraction.** Examples of these kinds of numbers are as follows:

(a) $\dfrac{\frac{2}{3}}{5}$ (b) $\dfrac{3}{\frac{4}{5}}$ (c) $\dfrac{\frac{2}{5}}{\frac{3}{7}}$

In each of the above, the longest line represents the main fraction line; the numerator is above this line, and the denominator is below it. Thus,

Rational Numbers 121

in (a) $\frac{2}{3}$ is the numerator of the compound fraction, and 5 is the denominator. In the hope that the following will clarify and not hopelessly confuse the issue, the 2 is the numerator of the numerator, and 3 is the denominator of the numerator, the 5 can be written as $\frac{5}{1}$ so that 5 is the numerator of the denominator, and finally, the 1 is the denominator of the denominator.

Compound fractions are more difficult to understand as well as to use in mathematical expressions. It is always desirable to simplify these kinds of fractions before attempting to work with them in any way.

This can easily be done by treating a compound fraction as the quotient of two rational numbers. Referring to examples (a), (b), and (c) again:

(a) $\dfrac{\frac{2}{3}}{5} = \dfrac{2}{3} \div 5 \qquad\qquad \dfrac{a}{b} = a \div b$

$\qquad\quad = \dfrac{2}{3} \cdot \dfrac{1}{5} \qquad\qquad a \div b = a \cdot \dfrac{1}{b}$

$\dfrac{\frac{2}{3}}{5} = \dfrac{2}{15} \qquad\qquad$ Multiplication of two rational numbers

(b) $\dfrac{\frac{3}{4}}{\frac{1}{5}} = 3 \div \dfrac{4}{5} \qquad\qquad \dfrac{a}{b} = a \div b$

$\qquad\quad = 3 \cdot \dfrac{5}{4} \qquad\qquad a \div b = a \cdot \dfrac{1}{b}$

$\qquad\quad = \dfrac{15}{4} \qquad\qquad a \cdot \dfrac{b}{c} = \dfrac{a \cdot b}{c}$

(c) $\dfrac{\frac{2}{5}}{\frac{3}{7}} = \dfrac{2}{5} \div \dfrac{3}{7} \qquad\qquad \dfrac{a}{b} = a \div b$

$\qquad\quad = \dfrac{2}{5} \cdot \dfrac{7}{3} \qquad\qquad a \div b = a \cdot \dfrac{1}{b}$

$\qquad\quad = \dfrac{14}{15} \qquad\qquad$ Multiplication of two rational numbers

Pre-Algebra

Generally then:

$$\frac{\frac{a}{b}}{c} = a \div \frac{b}{c} = a \cdot \frac{c}{b} = \frac{a \cdot c}{b}$$

$$\frac{\frac{a}{b}}{c} = \frac{a}{b} \div c = \frac{a}{b} \cdot \frac{1}{c} = \frac{a}{b \cdot c} \qquad \text{provided, in each case, that } b \neq 0, \\ c \neq 0, \text{ and } d \neq 0$$

$$\frac{\frac{a}{b}}{\frac{c}{d}} = \frac{a}{b} \div \frac{c}{d} = \frac{a}{b} \cdot \frac{d}{c} = \frac{a \cdot d}{b \cdot c}$$

Summarizing these results, compound fractions are those rational numbers which have fractions in the numerator, or denominator, or both. Compound fractions can be simplified by first treating them as division problems and then converting them into equivalent multiplication problems as previously defined.

3.9. Exercises

Simplify each of the following compound fractions into simple fractions which have no common factors:

1. $\dfrac{\frac{3}{7}}{4}$

2. $\dfrac{\frac{6}{5}}{9}$

3. $\dfrac{\frac{8}{5}}{12}$

4. $\dfrac{\frac{7}{5}}{9}$

5. $\dfrac{3}{\frac{7}{4}}$

6. $\dfrac{4}{\frac{8}{7}}$

7. $\dfrac{-8}{\frac{12}{5}}$

8. $\dfrac{-6}{\frac{5}{7}}$

9. $\dfrac{\frac{-5}{9}}{-3}$

10. $\dfrac{\frac{3}{8}}{-6}$

11. $\dfrac{\frac{-2}{-5}}{-6}$

12. $\dfrac{7}{\frac{8}{-3}}$

13. $\dfrac{\dfrac{-12}{5}}{4}$ 16. $\dfrac{-21}{\dfrac{-18}{7}}$ 19. $-\dfrac{\dfrac{-4}{7}}{\dfrac{16}{7}}$ 22. $\dfrac{\dfrac{-26}{8}}{\dfrac{-65}{24}}$

14. $\dfrac{\dfrac{4}{12}}{\dfrac{12}{5}}$ 17. $\dfrac{\dfrac{5}{8}}{\dfrac{3}{5}}$ 20. $\dfrac{\dfrac{3}{14}}{\dfrac{-8}{35}}$ 23. $\dfrac{\dfrac{-18}{56}}{\dfrac{54}{35}}$

15. $\dfrac{\dfrac{28}{-5}}{12}$ 18. $\dfrac{\dfrac{4}{7}}{\dfrac{3}{49}}$ 21. $-\dfrac{\dfrac{5}{8}}{\dfrac{-15}{32}}$ 24. $-\dfrac{\dfrac{-27}{12}}{\dfrac{12}{75}}$

3.10. Summary

A new kind of number has been defined and studied in this chapter. All pairs of integers which are written in the form $\dfrac{a}{b}$ are called rational numbers, providing that b is not zero. Every integer, a, can be written as $\dfrac{a}{1}$, so that every integer is a rational number, but not every rational number is an integer. Since the set of rational numbers contains the integers, the rational numbers must at least have the same properties as the integers. Accordingly, the following properties were found to apply for rational numbers:

> Multiplication of rational numbers is commutative and associative.
> Addition of rational numbers is commutative and associative.
> The identity element for addition is 0.
> The identity element for multiplication is 1.
> Each rational number has an additive inverse.
> Each nonzero rational number has a multiplicative inverse, or reciprocal.
> The distributive property holds for rational numbers.
> The rational numbers can be ordered.

Pre-Algebra

The set of integers has all these properties with the exception of one: possession of multiplicative inverse elements. The set of integers, and the natural numbers as well, also lacks the property of closure with respect to division. For this reason, division was not introduced until a set could be devised which was closed with respect to that operation. The set of rational numbers then is the "smallest" set which is closed for the arithmetic operations of addition, subtraction, multiplication, and division. In this chapter it was possible to discuss the division of integers and natural numbers, since both are subsets of the rational numbers. Division of zero by a nonzero number, division of a nonzero number by zero, and division of zero by zero were singled out as quotients which require particular attention.

The least common multiple (LCM), the least common denominator (LCD), and the greatest common factor (GCF), were concepts introduced here which are useful in the various operations involving rational numbers.

3.10. Review Exercises

Perform the indicated operations. Write your answer as a rational number with all common factors removed:

1. $8\left(\dfrac{3}{32}\right)$

2. $6\left(\dfrac{4}{36}\right)$

3. $-27\left(\dfrac{4}{81}\right)$

4. $-32\left(\dfrac{15}{28}\right)$

5. $\dfrac{8}{9} \cdot \dfrac{-5}{6}$

6. $\dfrac{-6}{27} \cdot \dfrac{-15}{14}$

7. $-\left(\dfrac{21}{15} \cdot \dfrac{-25}{13}\right)$

8. $-4\left(\dfrac{3}{5}\right) \cdot 7\left(\dfrac{30}{54}\right)$

9. $3{,}481 \cdot \left(\dfrac{583}{3{,}481}\right)$

10. $-8{,}031 \cdot \left(\dfrac{-164}{8{,}031}\right)$

11. $\dfrac{2}{7} \div 9$

12. $\dfrac{-10}{9} \div 25$

13. $-4 \div \dfrac{5}{6}$

14. $-14 \div \left(\dfrac{-8}{5}\right)$

15. $\dfrac{32}{17} \div \dfrac{-8}{5}$

16. $\dfrac{75}{-42} \div \dfrac{35}{-18}$

17. $6 \div 7 \div 8$

18. $-8 \div 6 \cdot 4$

19. $21 - 18 \div 3 - 2 \cdot 8$

20. $\dfrac{7}{3} \cdot (6 - 8) \div \dfrac{7}{12} (8 - 13)$

21. $\dfrac{8}{15} + \dfrac{2}{35}$

22. $\dfrac{2}{9} + \dfrac{4}{21}$

Pre-Algebra

23. $\dfrac{5}{8} - \dfrac{3}{14}$

24. $\dfrac{-7}{36} - \dfrac{7}{30}$

25. $5 \div 21 - 7 \div 30$

26. $4 \cdot \dfrac{5}{18} + 8 \div 5$

27. $7 - 4 \cdot 8 - 6(6 - 13) - (3 - 8)(8 - 3)$

28. $\dfrac{-7}{18} + \dfrac{23}{51}$

29. $8 - 17 \cdot 4 \div 12$
30. $13 \div 27 \cdot 18 - 9$
31. $7 \div 12 - 8 \div 21 - 9 \div 35$
32. $81 \div 13 \div 12 \div 18$

Write the reciprocals of each of the following:

33. $\dfrac{2}{3}$ 34. -5 35. $\dfrac{\frac{1}{4}}{7}$ 36. $\dfrac{-9}{5}$

Find the complete factorization for each of the following numbers:

37. 294   38. 770   39. 1,575   40. 864

Supply the missing numerator or denominator so that each of the following rational numbers are equal:

41. $\dfrac{8}{12} = \dfrac{}{-36}$ 43. $\dfrac{35}{-13} = \dfrac{-105}{}$

42. $\dfrac{-5}{8} = \dfrac{}{-48}$ 44. $\dfrac{-91}{-21} = \dfrac{}{-48}$

Rational Numbers 127

Find the LCM for each of the following sets of numbers:

45. {24, 30}
46. {28, 98}
47. {36, 210}
48. {42, 165}

49. {56, 147}
50. {75, 105}
51. {42, 63, 99}
52. {117, 156, 216}

Simplify each of the following compound fractions into simple fractions which have no common factors:

53. $\dfrac{\frac{4}{5}}{3}$

54. $\dfrac{\frac{-5}{9}}{15}$

55. $\dfrac{\frac{4}{5}}{\frac{5}{3}}$

56. $\dfrac{-8}{\frac{18}{5}}$

57. $\dfrac{\frac{-3}{7}}{\frac{5}{21}}$

58. $\dfrac{\frac{-5}{8}}{\frac{8}{-5}}$

Find the correct relationship (>, =, <) between each of the following pairs of numbers:

59. $\dfrac{3}{7}$ —— $\dfrac{4}{9}$

60. $\dfrac{4}{13}$ —— $\dfrac{16}{52}$

61. $\dfrac{-13}{7}$ —— $\dfrac{3}{2}$

62. $\dfrac{9}{5}$ —— $\dfrac{11}{6}$

4 Irrational Numbers

4.1. Exponents

In this chapter, an attempt shall be made to justify one more extension of the number system. Before doing so, it will be necessary to introduce several useful concepts.

One such concept is the use of **exponential notation** when writing the product of identical factors. For example:

(a) $3 \cdot 3 = 3^2$
(b) $4 \cdot 4 \cdot 4 = 4^3$
(c) $\dfrac{3}{4} \cdot \dfrac{3}{4} \cdot \dfrac{3}{4} \cdot \dfrac{3}{4} \cdot \dfrac{3}{4} = \left(\dfrac{3}{4}\right)^5$
(d) $10 \cdot 10 \cdot 10 \cdot 10 \cdot 10 \cdot 10 \cdot 10 \cdot 10 = 10^8$

Or more generally,

$$\underbrace{a \cdot a \cdot a \cdots a}_{n} = a^n \qquad \text{represents } n \text{ factors of } a, \text{ where } a \text{ is any rational number and } n \text{ is any natural number}$$

The 2, 3, 5, and 8 in examples (a), (b), (c), and (d), respectively, are each called **exponents.** The 3, 4, $\frac{3}{4}$, and 10 in examples (a), (b), (c), and (d), respectively, are each called the **base** of the exponent. In (a), 3^2 represents the product of 2 factors of 3, and is read as 3 to the second **power**, or more commonly, 3 squared. In (b), 4^3 represents the product of 3 factors of 4, and is read as 4 to the third power, or more commonly, 4 cubed. In (c), $(\frac{3}{4})^5$ represents the product of 5 factors of $\frac{3}{4}$, and is read as $\frac{3}{4}$ to the fifth power. When a number has been expressed with an exponent, the correct terminology is to say that it has been "raised to that power." However, numbers which are raised to the second and third powers are given the special names of square and cube. All others are read as "raised to the ... power," or more commonly, "to the ... power." The words "second," "third," "fourth," etc., are called **ordinal** numbers. Ordinal numbers represent the relative positions of objects in a given order, whereas, cardinal numbers represent different quantities of objects.

The negative of a number squared and the square of a negative number do not represent the same quantity. For example, -2^2 is the negative of two squared, or $-1 \cdot 2^2$, and $(-2)^2$ is the square of negative two. The first quantity, $-2^2 = -1 \cdot (2^2) = -4$, while the second quantity, $(-2)^2 = (-2)(-2) = +4$. Therefore, $-2^2 \neq (-2)^2$, and also $-2^4 \neq (-2)^4$, $-2^6 \neq (-2)^6$, etc.

The negative of a number cubed and the cube of a negative number do represent the same quantity. For example, -2^3 is the negative of two cubed, and $(-2)^3$ is the cube of negative two. The first quantity, $-2^3 = -1 \cdot (2^3) = -8$, while the second quantity, $(-2)^3 = (-2)(-2)(-2) = -8$. Therefore, $-2^3 = (-2)^3$, and also $-2^5 = (-2)^5$, $-2^7 = (-2)^7$, etc.

Integers which are multiples of 2 are called **even** numbers. The set of even numbers is $\{\ldots, -6, -4, -2, 0, 2, 4, 6, \ldots\}$. Integers which are not even are called **odd** numbers. The set of odd numbers is $\{\ldots, -7, -5, -3, -1, 1, 3, 5, 7, \ldots\}$. The above relationship can be generalized and then stated in terms of even and odd exponents, so that

$$-a^n = (-a)^n \quad \text{when } n \text{ is odd}$$

and

$$-a^n \neq (-a)^n \quad \text{when } n \text{ is even}$$

Irrational Numbers 131

Exponential notation is extremely useful when performing certain kinds of mathematical calculations. A few of these applications will be introduced here. For example, the product of numbers to the same base can be found in exponential notation.

(e) $\quad 3^2 \cdot 3^3 = (3 \cdot 3)(3 \cdot 3 \cdot 3)$
$\qquad\quad\;\; = (3 \cdot 3 \cdot 3 \cdot 3 \cdot 3) \qquad$ By the associative property
$\qquad\quad\;\; = 3^5$

(f) $\quad (-2)^3 \cdot (-2)^4 = [(-2)(-2)(-2)] \cdot [(-2)(-2)(-2)(-2)]$
$\qquad\qquad\qquad\;\; = [(-2)(-2)(-2)(-2)(-2)(-2)(-2)]$
$\qquad\qquad\qquad\;\; = (-2)^7$

(g) $\quad 10^4 \cdot 10^5 = (10 \cdot 10 \cdot 10 \cdot 10)(10 \cdot 10 \cdot 10 \cdot 10 \cdot 10)$
$\qquad\qquad\;\;\; = (10 \cdot 10 \cdot 10 \cdot 10 \cdot 10 \cdot 10 \cdot 10 \cdot 10 \cdot 10)$
$\qquad\qquad\;\;\; = 10^9$

Many more examples could be presented, but perhaps it is now obvious that

$$a^m \cdot a^n = a^{m+n} \qquad \text{where } a \text{ is any rational number, } m \text{ and } n \text{ any two natural numbers}$$

That is, the product of any two numbers in exponential notation, expressed to the same base, is that base expressed to the sum of the exponents. For example:

(h) $\;\; -7^2 \cdot 7^4 = -1 \cdot (7^2 \cdot 7^4) = -1 \cdot 7^6 = -7^6$
(i) $\;\; (-3)^4 \cdot (-3)^5 = (-3)^9$
(j) $\;\; \left(\dfrac{2}{3}\right)^3 \cdot \left(\dfrac{2}{3}\right) = \left(\dfrac{2}{3}\right)^4$

Note in (j) that although no exponent appears on the second factor, $\frac{2}{3}$ can be written as $(\frac{2}{3})^1$, so that $3 + 1 = 4$ is the sum of the exponents. Whenever there is only one factor of a given number, it can be expressed with an exponent of 1, although it usually is not. That is,

$$a^1 = a \qquad \text{for any rational number } a$$

Exponential notation is also useful to express any Hindu–Arabic numeral as the sum of multiples of powers of ten. The following examples will clarify this statement:

(k) $5{,}000 = 5 \cdot 1{,}000 = 5 \cdot 10 \cdot 10 \cdot 10 = 5 \cdot 10^3$
(l) $700 = 7 \cdot 100 = 7 \cdot 10 \cdot 10 = 7 \cdot 10^2$
(m) $30 = 3 \cdot 10 = 3 \cdot 10^1$

Some caution should be taken when reading or writing products of numbers involving exponents. For example, the exponent ^{3} in the product $5 \cdot 10^3$ affects only the base 10. In other words, $5 \cdot 10^3$ means the product of 1 factor of 5 and 3 factors of 10. More explicitly, only the base immediately next to the exponent is affected by the exponent, unless grouping symbols indicate otherwise.

In expanded notation, $5{,}734 = 5{,}000 + 700 + 30 + 4$, and with examples (k), (l), and (m), $5{,}734 = 5 \cdot 10^3 + 7 \cdot 10^2 + 3 \cdot 10^1 + 4$. Observe that the place value of each additional digit to the left in the Hindu–Arabic numeral system is represented by increasing the previous power of 10 by 1.

The quotient of numbers expressed in exponential notation, to the same base, can also be found. For example:

(n) $\dfrac{4^5}{4^2} = \dfrac{4 \cdot 4 \cdot 4 \cdot 4 \cdot 4}{4 \cdot 4}$

$= \dfrac{4 \cdot 4}{4 \cdot 4} \cdot \dfrac{4 \cdot 4 \cdot 4}{1}$

$= 1 \cdot (4 \cdot 4 \cdot 4)$

$= 4^3$

(o) $\dfrac{7^5}{7} = \dfrac{7 \cdot 7 \cdot 7 \cdot 7 \cdot 7}{7}$

$= \dfrac{7}{7} \cdot \dfrac{7 \cdot 7 \cdot 7 \cdot 7}{1}$

$= 1 \cdot (7 \cdot 7 \cdot 7 \cdot 7)$

$= 7^4$

Irrational Numbers   133

(p) $\dfrac{5^3}{5^5} = \dfrac{5 \cdot 5 \cdot 5}{5 \cdot 5 \cdot 5 \cdot 5 \cdot 5}$

$= \dfrac{5 \cdot 5 \cdot 5}{5 \cdot 5 \cdot 5} \cdot \dfrac{1}{5 \cdot 5}$

$= 1 \cdot \dfrac{1}{5^2}$

$= \dfrac{1}{5^2}$

(q) $\dfrac{10}{10^4} = \dfrac{10}{10 \cdot 10 \cdot 10 \cdot 10}$

$= \dfrac{10}{10} \cdot \dfrac{1}{10 \cdot 10 \cdot 10}$

$= 1 \cdot \dfrac{1}{10^3}$

$= \dfrac{1}{10^3}$

(r) $\dfrac{6^3}{6^3} = \dfrac{6 \cdot 6 \cdot 6}{6 \cdot 6 \cdot 6}$

$= \dfrac{6}{6} \cdot \dfrac{6}{6} \cdot \dfrac{6}{6}$

$= 1$

These examples illustrate that the quotient of numbers expressed in exponential notation with like bases fall into three different cases: one in which the exponent in the numerator exceeds that of the denominator; a second where the exponent in the numerator is less than that of the denominator, and a third where the exponents in the numerator and denominator are equal.

Note in (n) and (o), examples of the first case, that the exponent in the denominator is subtracted from that of the numerator. In (p) and (q), examples of the second case, observe that the exponent in the numerator is subtracted from the denominator exponent. Example (r) illustrates that when the exponents are equal, the value of the fraction is 1. These

Pre-Algebra

examples suggest the following set of relations:

$$\frac{a^m}{a^n} = a^{m-n} \quad \text{where } a \text{ is a nonzero rational number and } m > n$$

$$\frac{a^m}{a^n} = \frac{1}{a^{n-m}} \quad \text{where } a \text{ is a nonzero rational number and } m < n$$

$$\frac{a^m}{a^n} = 1 \quad \text{where } a \text{ is a nonzero rational number and } m = n$$

Using the above, $\frac{9^8}{9^3} = 9^5$, $\frac{6^2}{6^5} = \frac{1}{6^3}$, and $\frac{4^7}{4^7} = 1$.

Multiplication of rational numbers involving exponential notation can also be performed as demonstrated by the following examples:

(s) $\quad \dfrac{4^3}{7^5} \cdot \dfrac{4^2}{7^3} = \dfrac{4^3 \cdot 4^2}{7^5 \cdot 7^3} \qquad \dfrac{a}{b} \cdot \dfrac{c}{d} = \dfrac{a \cdot c}{b \cdot d}$

$\qquad \quad = \dfrac{4^5}{7^8} \qquad\qquad\qquad a^m \cdot a^n = a^{m+n}$

(t) $\quad \dfrac{6}{5^2} \cdot \dfrac{5^4}{6^5} = \dfrac{6 \cdot 5^4}{5^2 \cdot 6^5} \qquad \dfrac{a}{b} \cdot \dfrac{c}{d} = \dfrac{a \cdot c}{b \cdot d}$

$\qquad \quad = \dfrac{5^4 \cdot 6}{5^2 \cdot 6^5} \qquad\qquad$ Commutative property of multiplication

$\qquad \quad = \dfrac{5^4}{5^2} \cdot \dfrac{6}{6^5} \qquad\qquad \dfrac{a \cdot c}{b \cdot d} = \dfrac{a}{b} \cdot \dfrac{c}{d}$

$\qquad \quad = \dfrac{5^2}{1} \cdot \dfrac{1}{6^4}$

$\qquad \quad = \dfrac{5^2}{6^4}$

The results of example (t) indicate that the multiplication can be easily performed by finding the quotients $\dfrac{5^4}{5^2}$ and $\dfrac{6}{6^5}$ without any of the intermediate steps shown above.

Irrational Numbers   135

Multiplication and division of numbers expressed in exponential notation with like bases have been presented in this section. The following properties summarize these results:

$$a^m \cdot a^n = a^{m+n}$$

$$\frac{a^m}{a^n} = a^{m-n} \quad \text{when } m > n,\ a \neq 0$$

$$\frac{a^m}{a^n} = \frac{1}{a^{n-m}} \quad \text{when } m < n,\ a \neq 0$$

$$\frac{a^m}{a^n} = 1 \quad \text{when } m = n,\ a \neq 0$$

In each of these, a is a rational number, m and n are natural numbers.

4.1. Exercises

Write each of the following in exponential notation, writing each different numeral in your answer with a single exponent:

1. $2 \cdot 2 \cdot 2$

2. $(-5)(-5)(-5)(-5)$

3. $7 \cdot 7 \cdot 7 \cdot 4 \cdot 4 \cdot 4 \cdot 4$

4. $\frac{3}{5} \cdot \frac{3}{5} \cdot \frac{3}{5} \cdot \frac{3}{5} \cdot \frac{3}{5}$

5. $\frac{-4}{9} \cdot \frac{-4}{9} \cdot \frac{-4}{9} \cdot \frac{-4}{9}$

6. $-9 \cdot 9 \cdot 9 \cdot 9$

7. $(-3) \cdot 6 \cdot 6 \cdot (-3) \cdot (-3) \cdot 6 \cdot 6 \cdot 6$

8. $4(-2) \cdot 5(-2) \cdot 5 \cdot 5 \cdot 4 \cdot (-2) \cdot 5$

136   Pre-Algebra

Perform each of the indicated operations, writing each different numeral in your answer with a single exponent:

9. $3^5 \cdot 3^2 \cdot 3^4$

10. $(-2)^4 \cdot (-2)^3$

11. $5^3 \cdot 5^6$

12. $4^3 \cdot 4^5$

13. $\left(\dfrac{-2}{3}\right)^7 \cdot \left(\dfrac{-2}{3}\right)$

14. $\dfrac{4^3}{4}$

15. $\dfrac{5^7}{5^3}$

16. $\dfrac{(-7)^3}{(-7)^6}$

17. $\dfrac{-8^3}{8}$

18. $\dfrac{10}{10^6}$

19. $(-5^2)(5^7)$

20. $\dfrac{431^7}{431^7}$

21. $\dfrac{7^3 \cdot 7^8}{7^5 \cdot 7^6}$

22. $\dfrac{8^4 \cdot 8^5}{8^3}$

23. $\dfrac{5^3}{6^2} \cdot \dfrac{5^8}{6^3}$

24. $\dfrac{(-4)^6}{7^5} \cdot \dfrac{(-4)^4}{7^6}$

25. $\dfrac{8^4}{7^3} \cdot \dfrac{7^5}{8}$

26. $\dfrac{2^6}{3^4} \cdot \dfrac{3}{2^3}$

27. $\dfrac{-13^5}{13^9}$

28. $\dfrac{1}{2^4} \cdot \dfrac{-1}{2^7} \cdot \dfrac{1}{2^5}$

Find the integer or fraction which each of the following expressions in exponential notation represent:

29. $3^2 \cdot 4^2$ 30. $(-2)^3 \cdot 5^2$ 31. $(-4)^4$ 32. -4^4

Find the integer or fraction which each of the following expressions in exponential notation represent:

33. $\dfrac{3^2}{4^3}$

34. $\left(\dfrac{4}{5}\right)^2$

35. -3^5

36. $(-3)^5$

37. $-6^2 \cdot (-3)^2$

38. $-(-3)^3 \cdot (-2)^3$

39. $\left(\dfrac{-5}{6}\right)^3$

40. $\dfrac{(-2)^3}{(-3)^4}$

Irrational Numbers 137

Write each of the following numbers in expanded notation as multiples of powers of 10 (Example: $483 = 4 \cdot 10^2 + 8 \cdot 10 + 3$):

41. 834   42. 6,703   43. 48,689   44. 407,063

4.2. Operations with Exponents

A number already expressed in exponential notation can also be raised to a power. For example, $(3^2)^3$ means that there are three factors of 3^2, or $3^2 \cdot 3^2 \cdot 3^2$ which is equal to 3^6 so that $(3^2)^3 = 3^6$. Again, $(5^2)^4 = 5^2 \cdot 5^2 \cdot 5^2 \cdot 5^2$, or equivalently, 5^8, so that $(5^2)^4 = 5^8$. Perhaps the pattern is now clear that when raising numbers written in exponential notation to a power, the exponents are multiplied. In mathematical symbolism then, this becomes:

$(a^m)^n = a^{m \cdot n}$ where a is a rational number, m and n are natural numbers

All factors inside a grouping symbol are affected by a given exponent. For example:

(a) $(5 \cdot 3)^2 = (5 \cdot 3)(5 \cdot 3)$
$= 5 \cdot 5 \cdot 3 \cdot 3$ By the associative and
$= 5^2 \cdot 3^2$ commutative properties

(b) $(4 \cdot 5^2)^3 = (4 \cdot 5^2)(4 \cdot 5^2)(4 \cdot 5^2)$
$= (4 \cdot 4 \cdot 4)(5^2 \cdot 5^2 \cdot 5^2)$ By associative and commuta-
$= 4^3 \cdot 5^6$ tive properties

Without resorting to additional examples, perhaps it is now obvious that raising the product of numbers to a given power raises all the factors of the product to that power.
Symbolically then:

$(a \cdot b)^n = a^n \cdot b^n$ where a and b are rational numbers, and n is a natural number

Additional examples illustrating this relationship are:

$(3 \cdot 6^4)^2 = 3^2 \cdot 6^8$ and $(5^2 \cdot 4^3)^4 = 5^8 \cdot 4^{12}$

Rational numbers raised to a given power can also be found in exponential notation by a simple relation. The following examples illustrate these kinds of problems:

(c) $\left(\dfrac{5}{7}\right)^3 = \dfrac{5}{7} \cdot \dfrac{5}{7} \cdot \dfrac{5}{7}$

$= \dfrac{5 \cdot 5 \cdot 5}{7 \cdot 7 \cdot 7}$

$= \dfrac{5^3}{7^3}$

(d) $\left(\dfrac{3^5}{4^3}\right)^2 = \dfrac{3^5}{4^3} \cdot \dfrac{3^5}{4^3}$

$= \dfrac{(3^5)^2}{(4^3)^2}$

$= \dfrac{3^{10}}{4^6}$

Note in examples (c) and (d) that when raising these rational numbers to a given power, both the numerator and denominator are raised to that power. This can be stated symbolically as follows:

$\left(\dfrac{a}{b}\right)^n = \dfrac{a^n}{b^n}$ where a and b are rational numbers, $b \neq 0$ and n is a natural number

The following properties of exponents summarize the results of this and the previous section:

$$a^m \cdot a^n = a^{m+n}$$

$$\dfrac{a^m}{a^n} = a^{m-n} \qquad \text{when } m > n, a \neq 0$$

$$\dfrac{a^m}{a^n} = \dfrac{1}{a^{n-m}} \qquad \text{when } m < n, a \neq 0$$

$$\dfrac{a^m}{a^n} = 1 \qquad \text{when } m = n, a \neq 0$$

$$(a^m)^n = a^{m \cdot n}$$
$$(a \cdot b)^n = a^n \cdot b^n$$
$$\left(\dfrac{a}{b}\right)^n = \dfrac{a^n}{b^n} \qquad \text{provided that } b \neq 0$$

Irrational Numbers 139

In each of these, a and b are rational numbers, m and n are natural numbers.

4.2. Exercises

Perform each of the indicated operations, writing each numeral in your answer with a single exponent:

1. $(5^2)^3$
2. $5^2 \cdot 5^3$
3. $(-8^3)^4$
4. $-8^3 \cdot 8^4$
5. $(-7^5)^3$
6. $[(-3)^4]^6$
7. $(5 \cdot 7)^3$
8. $[(-3) \cdot 4]^4$
9. $(3^2 \cdot 8)^6$
10. $(4^3 \cdot 5^2)^5$
11. $(3^2)^3 \cdot (3^4)^2$
12. $(-4^2)^3 \cdot (-4^3)^4$
13. $(2^3 \cdot 3^4 \cdot 4^5)^6$
14. $(-3 \cdot 4^3)^4$
15. $\left(\dfrac{3}{4^2}\right)^4$
16. $\left(\dfrac{5^3}{4^5}\right)^3$
17. $\left(\dfrac{-6}{7^4}\right)^6$
18. $\left(\dfrac{2 \cdot 3^4}{5 \cdot 6^3}\right)^3$
19. $\left(\dfrac{8^3 \cdot 8^6}{8^5 \cdot 8^9}\right)^3$
20. $\left(\dfrac{5 \cdot 7^3}{7^8}\right)^4$
21. $[2 \cdot (3^2 \cdot 4^3)^5]^3$
22. $\dfrac{(9^3)^5}{(9^5)^2}$
23. $\dfrac{(-3^4)^2}{(-3^7)^3}$
24. $\dfrac{7^3 \cdot (-7)^5}{(-7)^4 \cdot 7^8}$

4.3. Radicals

It is possible to raise any rational number to a given power, provided that power is a natural number. The quantities involved may become large, but they can always be found. The reverse problem is more difficult and not always solvable in the system of rational numbers. This is to express any rational number as the power of a given number. For example:

(a) $4 = (2)^2$ and $4 = (-2)^2$
(b) $9 = (3)^2$ and $9 = (-3)^2$
(c) $-8 = (-2)^3$
(d) $27 = (3)^3$

The problem in (a), was to write 4 as the square of some number. Observe that there are two numbers, $+2$ and -2, which provide the desired result. Both $+2$ and -2 are called the **square roots** of 4.

> The square root of any number is defined as one of the two equal factors of that number.

The product of two factors of 4 and -4 each produce 16, so that 4 and -4 are the two square roots of 16. The square roots of 16 are written $+\sqrt{16}$ and $-\sqrt{16}$, so that $+\sqrt{16} = +4$ and $-\sqrt{16} = -4$. The positive square root of a number is called the **principal** square root, so that $+\sqrt{16} = \sqrt{16}$ is the principal square root of 16. The $\sqrt{}$ sign is called a **radical sign,** and the number inside the radical sign is called the **radicand,** In (b), there are also two square roots of 9, namely, $+3$ and -3, so that $\sqrt{9} = 3$ and $-\sqrt{9} = -3$. Since each positive number is either the product of two positive factors or two negative factors, each positive number has two square roots. In general:

> If $b^2 = a$, then $b = \sqrt{a}$ or $b = -\sqrt{a}$, providing that $a \geq 0$.

To illustrate, if $b^2 = 36$, then $b = \sqrt{36}$, or $b = -\sqrt{36}$. The radical expressions $\sqrt{25}$ and $-\sqrt{49}$ each indicate one particular square root of 25 and 49, namely, 5 and -7, respectively. Although each positive number has two square roots, each radical expressions can have one and only one numerical value. Please note that the square root of a nonzero number is always positive. The value of a radical can only be negative when a negative sign is in front of the radical. Although $\sqrt{a}$ can never be negative, the value of $\sqrt{a}$ can be zero when $a = 0$. That is, $\sqrt{0} = 0$ since one of the two equal factors of 0 is 0.

When finding the square root of 4, a mathematical process is performed on the number 4, so that the number 2 is obtained. Here then is an operation performed on one number to produce a second number. For that reason, the process of finding the root of a number is called a **unary operation.** Squaring a positive number and then finding its square root are inverse operations, but this relationship does not hold for negative numbers. Squaring -4 produces 16, but $\sqrt{16} = +4$, which is not the original number. To illustrate:

$$\sqrt{2^2} = \sqrt{4} = 2$$
$$\sqrt{4^2} = \sqrt{16} = 4$$
$$\sqrt{(-3)^2} = \sqrt{9} = 3$$

Restating this idea, the square root of any squared number is always nonnegative. Symbolically, then:

$$\sqrt{a^2} = |a| \quad \text{where } a \text{ is any rational number}$$

Using this principle, $\sqrt{169^2} = 169$ and $\sqrt{(-4{,}376)^2} = 4{,}376$.

In (c), since the cube of -2 is -8, -2 is called the **cube root** of -8. In (d), 3 is the cube root of 27. These are written symbolically as follows:

(e) $\sqrt[3]{-8} = -2$
(f) $\sqrt[3]{27} = 3$

Observe that

$$\sqrt[3]{-8} = \sqrt[3]{(-2)^3} = -2$$

and

$$\sqrt[3]{27} = \sqrt[3]{(3)^3} = 3$$

Here then, the successive operations of cubing and cube root result in the given number. Cubing any number and then taking the cube root of the product are inverse operations.

The number on the radical sign, in this case 3, is called the **index.** Note that the index is not written when finding the square root. With the use of the index, any root of a number can be indicated. Again, a few examples may make this concept clear:

(g) Since $(-2)^5 = -32$, then $\sqrt[5]{-32} = -2$.
(h) Since $3^4 = 81$, then $\sqrt[4]{81} = 3$.

In general:

$$\text{If} \quad b^n = a, \quad \text{then} \quad \sqrt[n]{a} = b.$$

When n is an even natural number, a is restricted to a nonnegative rational number. When n is an odd natural number, a can be any rational number. The odd root of a negative number is found in example (g) above. Although the fourth root of 81 is found in (h), the fourth root of -81 is undefined since no number exists whose fourth power is -81.

The term "radical" includes the entire expression: radical sign, radicand, and index, when necessary. Therefore, $\sqrt{6}$ and $\sqrt[3]{2}$ are examples of two radicals.

Whenever a number has two equal factors which are rational numbers, that number is said to be a **perfect square.** Therefore, 1, 4, 9, 16, and 25 are perfect squares. Numbers such as 8, 27, 64, and 125, which are products of three equal factors are referred to as **perfect cubes.** Any number which is the product of a given number of identical factors is called a **perfect power.** For example, 32 is a perfect fifth power.

A number such as 2 is not a perfect power, and in particular, it is not a perfect square. That is, there is no rational number which is one of the two equal factors of 2. Its principal square root, $\sqrt{2}$, can be indicated, but it has no rational number to which it is equivalent. Numbers such as $\sqrt{2}$, $\sqrt{3}$, $\sqrt{5}$, $\sqrt{6}$, also fall into this category and are called **irrational** numbers. Generally then,

> An irrational number is any number which cannot be represented as the product of a desired number of equal rational factors.

That is, an irrational number is one whose square root cannot be written as the product of two equal rational factors, or its cube root cannot be written as the product of three equal rational factors, etc. The set of irrational numbers will be designated by the letter H.

Irrational numbers have no exact rational value, but rational approximations can be found with as high a degree of accuracy as desired.

For example, it is clear that

$$\frac{196}{100} < \frac{200}{100} < \frac{225}{100}$$

which can be written as

$$\left(\frac{14}{10}\right)^2 < 2 < \left(\frac{15}{10}\right)^2$$

Irrational Numbers   143

and the square roots of these are

$$\frac{14}{10} < \sqrt{2} < \frac{15}{10}$$

So that $\sqrt{2}$ is a number between $\frac{14}{10}$ and $\frac{15}{10}$.

For a closer approximation, it should be clear that

$$\frac{19{,}881}{10{,}000} < \frac{20{,}000}{10{,}000} < \frac{20{,}164}{10{,}000}$$

which can be written as

$$\left(\frac{141}{100}\right)^2 < 2 < \left(\frac{142}{100}\right)^2$$

and the square roots of these are

$$\frac{141}{100} < \sqrt{2} < \frac{142}{100}$$

So that $\sqrt{2}$ is between $\frac{141}{100}$ and $\frac{142}{100}$.

Closer approximations can be found, but these examples should be sufficient at this time to justify the existence of the irrational number $\sqrt{2}$. A much improved method of representing these approximations will be presented in Chapter 5.

The product of radicals can also be found. For example, since $\sqrt{4} = 2$, then $\sqrt{4} \cdot \sqrt{4} = 2 \cdot 2$, or $\sqrt{4} \cdot \sqrt{4} = 4$. Likewise, $\sqrt{25} = 5$, and $\sqrt{25} \cdot \sqrt{25} = 5 \cdot 5$, so that $\sqrt{25} \cdot \sqrt{25} = 25$. In this manner, it should be easy to see that

$$\sqrt{a} \cdot \sqrt{a} = a \quad \text{where } a \text{ is any nonnegative rational number}$$

Since $\sqrt{a} \cdot \sqrt{a} = (\sqrt{a})^2$, it is also true that

$$(\sqrt{a})^2 = a \quad \text{where } a \text{ is any nonnegative rational number}$$

The following examples lead to an additional result which is quite useful:

(i) $\sqrt{4} \cdot \sqrt{4} = 2 \cdot 2$
$\phantom{\sqrt{4} \cdot \sqrt{4}} = 4$
$\sqrt{4} \cdot \sqrt{4} = \sqrt{16} = \sqrt{4 \cdot 4}$ \quad\quad Since $4 = \sqrt{16}$

Pre-Algebra

(j) $(\sqrt{4} \cdot \sqrt{2})^2 = (\sqrt{4})^2 \cdot (\sqrt{2})^2$ $(a \cdot b)^2 = a^2 \cdot b^2$
$\phantom{(\sqrt{4} \cdot \sqrt{2})^2} = 4 \cdot 2$ $(\sqrt{a})^2 = a$
$(\sqrt{4} \cdot \sqrt{2})^2 = 8$

Therefore,
$$\sqrt{4} \cdot \sqrt{2} = \sqrt{8} \qquad \text{If } b^2 = a, \text{ then } b = \sqrt{a}$$

and also,
$$\sqrt{4} \cdot \sqrt{2} = \sqrt{4 \cdot 2} \qquad \text{Since } \sqrt{8} = \sqrt{4 \cdot 2}$$

(k) $(\sqrt{4} \cdot \sqrt{3})^2 = (\sqrt{4})^2 \cdot (\sqrt{3})^2$ $(a \cdot b)^2 = a^2 \cdot b^2$
$\phantom{(\sqrt{4} \cdot \sqrt{3})^2} = 4 \cdot 3$
$\phantom{(\sqrt{4} \cdot \sqrt{3})^2} = 12$

Therefore,
$$\sqrt{4} \cdot \sqrt{3} = \sqrt{12} = \sqrt{4 \cdot 3} \qquad \text{If } b^2 = a, \text{ then } b = \sqrt{a}$$

Note the results in the last three examples:

$$\sqrt{4 \cdot 4} = \sqrt{4} \cdot \sqrt{4} \qquad \sqrt{4 \cdot 2} = \sqrt{4} \cdot \sqrt{2} \qquad \sqrt{4 \cdot 3} = \sqrt{4} \cdot \sqrt{3}$$

This pattern can be verbalized by the statement:

> The square root of the product of two nonnegative numbers is equal to the product of their square roots.

Which can be written symbolically as:

$$\sqrt{a \cdot b} = \sqrt{a} \cdot \sqrt{b} \qquad \text{where } a \text{ and } b \text{ are any two nonnegative rational numbers}$$

Note that the statement above is restricted to the use of nonnegative numbers only. The following examples demonstrate that if a or b, or both, are negative, the statement is not valid:

(l) $\sqrt{(-2)(-4)} \neq \sqrt{-2} \cdot \sqrt{-4}$ Since neither $\sqrt{-2}$ nor $\sqrt{-4}$ are defined

(m) $\sqrt{2(-4)} \neq \sqrt{2} \cdot \sqrt{-4}$ Since neither $\sqrt{2(-4)}$ nor $\sqrt{-4}$ are defined

Irrational Numbers   145

Radicals can now be simplified by removing any perfect squares which are contained in the radicand. For example:

(n) $\sqrt{8} = \sqrt{4 \cdot 2}$
$= \sqrt{4} \cdot \sqrt{2}$
$= 2 \cdot \sqrt{2}$ or simply $2\sqrt{2}$

(o) $\sqrt{12} = \sqrt{4 \cdot 3}$
$= \sqrt{4} \cdot \sqrt{3}$
$= 2 \cdot \sqrt{3}$ or simply $2\sqrt{3}$

(p) $\sqrt{24} = \sqrt{4 \cdot 6}$
$= \sqrt{4} \cdot \sqrt{6}$
$= 2\sqrt{6}$

Now observe that $\sqrt[3]{8} = 2$, so that $\sqrt[3]{8} \cdot \sqrt[3]{8} \cdot \sqrt[3]{8} = 2 \cdot 2 \cdot 2$, and therefore, $(\sqrt[3]{8})^3 = 8$. In a similar manner, $\sqrt[3]{-27} = -3$, so that $(\sqrt[3]{-27})(\sqrt[3]{-27})(\sqrt[3]{-27}) = -3 \cdot (-3)(-3)$, and therefore, $(\sqrt[3]{-27})^3 = -27$. Generally then:

$$(\sqrt[3]{a})(\sqrt[3]{a})(\sqrt[3]{a}) = a \quad \text{for any rational number } a$$

and since $(\sqrt[3]{a})(\sqrt[3]{a})(\sqrt[3]{a}) = (\sqrt[3]{a})^3$, then

$$(\sqrt[3]{a})^3 = a \quad \text{for any rational number } a$$

It is generally true that

$$(\sqrt[n]{a})^n = a \quad \text{for all } a \text{ and } n \text{ for which } \sqrt[n]{a} \text{ is defined}$$

Now observe the following examples:

(q) $(\sqrt[3]{8} \cdot \sqrt[3]{2})^3 = (\sqrt[3]{8})^3 \cdot (\sqrt[3]{2})^3$ $(a \cdot b)^3 = a^3 \cdot b^3$
$= 8 \cdot 2$
$= 16$

Therefore,

$\sqrt[3]{8} \cdot \sqrt[3]{2} = \sqrt[3]{16} = \sqrt[3]{8 \cdot 2}$ If $b^3 = a$, then $b = \sqrt[3]{a}$

(r) $(\sqrt[3]{8} \cdot \sqrt[3]{3})^3 = (\sqrt[3]{8})^3 \cdot (\sqrt[3]{3})^3$ $(a \cdot b)^3 = a^3 \cdot b^3$
$= 8 \cdot 3$
$= 24$

Therefore,

$$\sqrt[3]{8} \cdot \sqrt[3]{3} = \sqrt[3]{24} = \sqrt[3]{8 \cdot 3} \qquad \text{If } b^3 = a, \text{ then } b = \sqrt[3]{a}$$

Examples (q) and (r) demonstrate that

$$\sqrt[3]{8 \cdot 2} = \sqrt[3]{8} \cdot \sqrt[3]{2} \quad \text{and} \quad \sqrt[3]{24} = \sqrt[3]{8} \cdot \sqrt[3]{3}$$

These examples suggest that $\sqrt[3]{a} \cdot \sqrt[3]{b} = \sqrt[3]{a \cdot b}$. It was previously determined that $\sqrt{a}\sqrt{b} = \sqrt{a \cdot b}$, and it can be established generally that:

$$\sqrt[n]{a \cdot b} = \sqrt[n]{a} \cdot \sqrt[n]{b} \qquad \begin{array}{l} n \text{ a natural number such that } n \geq 3 \text{ when } n \text{ is} \\ \text{odd, } a \text{ and } b \text{ are any two rational numbers} \\ \text{when } n \text{ is even, } a \text{ and } b \text{ are two nonnegative} \\ \text{rational numbers} \end{array}$$

Radicals which contain perfect cube roots can now be simplified using the above principle as follows:

(s) $\quad \sqrt[3]{54} = \sqrt[3]{27 \cdot 2}$
$\qquad\quad = \sqrt[3]{27} \cdot \sqrt[3]{2}$
$\qquad\quad = 3 \cdot \sqrt[3]{2} \qquad$ or simply $\quad 3\sqrt[3]{2}$

(t) $\quad \sqrt[3]{48} = \sqrt[3]{8 \cdot 6}$
$\qquad\quad = \sqrt[3]{8} \cdot \sqrt[3]{6}$
$\qquad\quad = 2 \cdot \sqrt[3]{6} \qquad$ or simply $\quad 2\sqrt[3]{6}$

Observe that the multiplication dot is not necessary in order to indicate multiplication.

Finding the root of a number, which is a unary operation, has been discussed here. Radicals were introduced and some of the properties of this convenient notation were developed. The power of a number and its root are seen to be inverse operations under certain conditions. Of more importance is the concept that:

$$\text{If} \quad b^n = a, \quad \text{then} \quad b = \sqrt[n]{a}.$$

If a is the perfect nth power of b, then b is a rational number. If a is not the perfect nth power of b, then b is an irrational number. A number cannot be both rational and irrational so that the new set of irrational

Irrational Numbers 147

numbers, H, is disjoint from the set of rational numbers, Q. Since the sum or difference of two irrational numbers may be the rational number zero, H is not closed with respect to addition or subtraction. The product or quotient of two irrational numbers may also be rational and therefore, H is not closed with respect to multiplication or division. In particular, $\sqrt{2} \cdot \sqrt{2} = 2$ and $\dfrac{\sqrt{2}}{\sqrt{2}} = 1$.

4.3. Exercises

Find the square roots of each of the given numbers:
1. 81
2. 64
3. 121
4. 36
5. 1
6. 10,000
7. 225
8. 144

Simplify each of the following radicals by finding a rational number or removing any perfect power from the radicand:

9. $\sqrt{25}$
10. $-\sqrt{49}$
11. $\sqrt[3]{-8}$
12. $-\sqrt[3]{125}$
13. $-\sqrt{36}$
14. $\sqrt[4]{16}$
15. $\sqrt[3]{-216}$
16. $-\sqrt{169}$

Find the product of each of the following. Express your answer as an integer or rational number whenever possible:

17. $\sqrt{2} \cdot \sqrt{2}$
18. $\sqrt{7} \cdot \sqrt{7}$
19. $\sqrt{2} \cdot \sqrt{8}$
20. $\sqrt{473} \cdot \sqrt{473}$
21. $\sqrt[3]{2} \cdot \sqrt[3]{4}$
22. $\sqrt[3]{3} \cdot \sqrt[3]{3}$
23. $\sqrt[4]{4} \cdot \sqrt[4]{4}$
24. $\sqrt[3]{4} \cdot \sqrt[3]{16}$
25. $\sqrt{2}\sqrt{3}\sqrt{6}$
26. $\sqrt{3}\sqrt{4}\sqrt{12}$
27. $\sqrt[3]{4} \cdot \sqrt[3]{9} \cdot \sqrt[3]{6}$
28. $\sqrt[3]{(-7)^3}$
29. $\sqrt{(-7)^2}$
30. $(\sqrt{47})^2$
31. $(\sqrt[3]{17})^3$
32. $(\sqrt[5]{47})^5$

Simplify each of the following radicals by removing all factors which are perfect powers from the radicand:

33. $\sqrt{18}$
34. $\sqrt{45}$
35. $\sqrt{72}$
36. $\sqrt[3]{32}$
37. $\sqrt[3]{-81}$
38. $\sqrt{27}$
39. $\sqrt{48}$
40. $\sqrt[4]{32}$
41. $\sqrt{32}$
42. $\sqrt[3]{72}$
43. $\sqrt{75}$
44. $\sqrt{98}$
45. $\sqrt{54}$
46. $\sqrt[3]{54}$
47. $\sqrt[3]{16}$
48. $\sqrt{63}$

4.4. Operations with Radicals

Calculations in algebra, and other advanced mathematics courses, often involve addition, subtraction, multiplication, and division of radicals. It would be well to become familiar with these operations using radicals, at this time, so that there will be less difficulty with these ideas in later courses.

Radicals with like radicands and indices (plural of index), are called **like radicals.** If two radicals have unlike indices, unlike radicands, or both, they are said to be **unlike radicals.** Like radicals can be added as shown in the following examples:

(a) $\sqrt{2} + \sqrt{2} + \sqrt{2}$
$= 1 \cdot \sqrt{2} + 1 \cdot \sqrt{2} + 1 \cdot \sqrt{2}$ $\qquad a = 1 \cdot a$
$= (1 + 1 + 1) \cdot \sqrt{2}$ $\qquad a \cdot c + b \cdot c = (a + b) \cdot c$
$= 3\sqrt{2}$

(b) $\sqrt[3]{5} + \sqrt[3]{5} = 1 \cdot \sqrt[3]{5} + 1 \cdot \sqrt[3]{5}$ $\qquad a = 1 \cdot a$
$= (1 + 1)\sqrt[3]{5}$ $\qquad a \cdot c + b \cdot c = (a + b) \cdot c$
$= 2\sqrt[3]{5}$

From (a), $3\sqrt{2}$ can be interpreted as meaning the product of 3 and $\sqrt{2}$, or the sum of three $\sqrt{2}$'s. A similar interpretation can be made in (b), and these can be generalized so that

$a\sqrt{b}$ is defined either as the product of a and $\sqrt{b}$ or as the sum
$$\underbrace{\sqrt{b} + \sqrt{b} + \cdots + \sqrt{b}}_{a \text{ terms}}$$

To illustrate, $6\sqrt{3}$, represents the product of 6 and $\sqrt{3}$ or the sum of six $\sqrt{3}$'s. Likewise, the quantity $-3\sqrt{5}$ is the negative of the product of 3 and $\sqrt{5}$, or the opposite of the sum of three $\sqrt{5}$'s.

Radicals with unlike radicands cannot be added or subtracted. That is, $\sqrt{6} + \sqrt{3}$, cannot be represented by a single radical. In particular, $\sqrt{6} + \sqrt{3} \neq \sqrt{9}$, and in general,

$\qquad \sqrt{a} + \sqrt{b} \neq \sqrt{a + b} \qquad$ when a and b are nonzero rational numbers

Irrational Numbers 149

Two radicals which are unlike can be combined if perfect squares can be removed from either one or both so that the resulting radicals are alike. The two expressions, $\sqrt{3}$ and $-4\sqrt{3}$, are considered to be like radicals even though they contain unlike factors. For example:

(c) $\sqrt{8} + \sqrt{2} = \sqrt{4} \cdot \sqrt{2} + \sqrt{2}$
$= 2\sqrt{2} + 1 \cdot \sqrt{2}$
$= 3\sqrt{2}$

(d) $\sqrt{12} + \sqrt{27} = \sqrt{4} \cdot \sqrt{3} + \sqrt{9} \cdot \sqrt{3}$
$= 2\sqrt{3} + 3\sqrt{3}$
$= 5\sqrt{3}$

(e) $\sqrt{18} + 2\sqrt{8} - 3\sqrt{50} = \sqrt{9} \cdot \sqrt{2} + 2(\sqrt{4} \cdot \sqrt{2}) - 3(\sqrt{25} \cdot \sqrt{2})$
$= 3\sqrt{2} + 2(2\sqrt{2}) - 3(5\sqrt{2})$
$= 3\sqrt{2} + 4\sqrt{2} - 15\sqrt{2}$
$= (3 + 4 - 15) \cdot \sqrt{2}$
$= -8\sqrt{2}$

(f) $\sqrt{48} - \sqrt{12} + \sqrt{8} = \sqrt{16} \cdot \sqrt{3} - \sqrt{4} \cdot \sqrt{3} + \sqrt{4} \cdot \sqrt{2}$
$= 4\sqrt{3} - 2\sqrt{3} + 2\sqrt{2}$
$= 2\sqrt{3} + 2\sqrt{2}$

Example (f) cannot be simplified any further, since the two radicands are unlike.

The numerator or denominator of fractions, or both, are frequently radicals. Radicals should always be simplified by removing any perfect square which may be in the radicand. If the fraction then has a common factor in the numerator and denominator, it should be removed by using the fundamental property of fractions. The following example illustrates a calculation of this kind. The next section will include a discussion of the case in which a radical occurs in the denominator of a fraction, or in both terms of the fraction.

150   Pre-Algebra

(g) $\dfrac{\sqrt{18}}{6} = \dfrac{\sqrt{9} \cdot \sqrt{2}}{6}$

$= \dfrac{3\sqrt{2}}{3 \cdot 2}$

$= \dfrac{\sqrt{2}}{2}$

Expressions involving radicals can also be raised to a power. The resulting expression should be simplified whenever possible by removing any perfect square from the radicand.

The following examples will illustrate this kind of calculation:

(h) $(4\sqrt{2})^2 = (4\sqrt{2})(4\sqrt{2})$
$= 4 \cdot 4 \cdot \sqrt{2} \cdot \sqrt{2}$
$= 16 \cdot 2$
$= 32$

(i) $(-3\sqrt{5})^3 = (-3\sqrt{5})(-3\sqrt{5})(-3\sqrt{5})$
$= (-3)(-3)(-3)[(\sqrt{5})(\sqrt{5})](\sqrt{5})$
$= -27 \cdot 5\sqrt{5}$
$= -135\sqrt{5}$

Examples (h) and (i) can also be simplified more directly by using the property, $(a \cdot b)^n = a^n \cdot b^n$. In (h) then, $(4\sqrt{2})^2 = 4^2(\sqrt{2})^2 = 16 \cdot 2 = 32$. In (i), $(-3\sqrt{5})^3 = (-3)^3 \cdot (\sqrt{5})^3 = -27 \cdot (\sqrt{5})^2 \cdot \sqrt{5} = -27 \cdot 5 \cdot \sqrt{5} = -135\sqrt{5}$.

Quite frequently, it is necessary to multiply a radical by the sum of two radicals or by the sum of a rational number and a radical. The following multiplications were accomplished by using the distributive property, and then simplifying the radicals.

(j) $\sqrt{2}(3 + \sqrt{6}) = 3\sqrt{2} + \sqrt{2} \cdot \sqrt{6}$ Distributive property
$= 3\sqrt{2} + \sqrt{2}(\sqrt{2} \cdot \sqrt{3})$ $\sqrt{6} = \sqrt{2} \cdot \sqrt{3}$
$= 3\sqrt{2} + 2\sqrt{3}$

Irrational Numbers

(k) $4\sqrt{3}(\sqrt{3} - 6\sqrt{18}) = 4\sqrt{3}\sqrt{3} - (4\sqrt{3})(6\sqrt{18})$
$= 4 \cdot 3 - 4 \cdot 6 \cdot \sqrt{3}\sqrt{9}\sqrt{2}$
$= 12 - 24 \cdot 3 \cdot \sqrt{3}\sqrt{2}$
$= 12 - 72\sqrt{6}$

Note in (j) that these radicals cannot be combined since they are unlike. In (k), multiplication must be performed before subtraction, so that the 12 and (-72) cannot be combined.

In this section, radicals were added, subtracted, and multiplied. Any two radicals can be multiplied, providing their indices are the same. Only like radicals can be added or subtracted. Sometimes unlike radical expressions can be made into like radicals and then combined, by removing perfect squares from their radicands.

4.4. Exercises

Write each of the following sums or differences as a product or single fraction whenever possible:

1. $\sqrt{3} + \sqrt{3} + \sqrt{3} + \sqrt{3}$
2. $\sqrt{7} + \sqrt{5} + \sqrt{7} + \sqrt{5} + \sqrt{5}$
3. $2\sqrt{6} - 5\sqrt{6} - 9\sqrt{6}$
4. $\sqrt[3]{2} + 2\sqrt[3]{2} - 3\sqrt[3]{2}$
5. $-\sqrt{3} - \sqrt{3} - \sqrt{3} - \sqrt{2}$
6. $3\sqrt[3]{4} + 2\sqrt[3]{3} + 3\sqrt[3]{2}$
7. $5\sqrt[3]{6} + 3\sqrt[3]{6} - 4\sqrt{6}$
8. $2\sqrt[3]{3} + 3\sqrt[3]{5} - \sqrt[5]{3} - 4\sqrt[3]{5}$

Simplify each of the following radicals by removing all perfect squares from the radicand:

9. $3\sqrt{8}$
10. $-4\sqrt{18}$
11. $-6\sqrt{98}$
12. $\dfrac{\sqrt{28}}{4}$
13. $\dfrac{\sqrt{12}}{4}$
14. $\dfrac{-2\sqrt{27}}{3}$
15. $\dfrac{-\sqrt{32}}{6}$
16. $\dfrac{-2\sqrt{45}}{6}$
17. $\dfrac{\sqrt{75}}{10}$
18. $\dfrac{3\sqrt{40}}{8}$
19. $3\sqrt{24}$
20. $\dfrac{\sqrt[3]{108}}{9}$

Write each of the following products or powers as simplified radicals or integers:

21. $(\sqrt{7})^2$
22. $(\sqrt{5})^3$
23. $(3\sqrt{2})^2$
24. $\sqrt{6}\sqrt{2}$
25. $\sqrt{8}\sqrt{3}$
26. $(-4\sqrt{3})^2$
27. $\sqrt{6}\sqrt{3}\sqrt{8}$
28. $(\sqrt{3})^4$
29. $(\sqrt{2})^5$
30. $(-2\sqrt{3})^3$

31. $\sqrt{2}(\sqrt{6} - 3)$
32. $\sqrt{8}(\sqrt{2} - \sqrt{3})$
33. $\sqrt{6}(\sqrt{6} - \sqrt{8})$
34. $3\sqrt{2}(3\sqrt{2} + 4\sqrt{6})$
35. $-4\sqrt{3}(\sqrt{27} - 3\sqrt{5})$
36. $\sqrt[3]{2}(\sqrt[3]{4} + \sqrt[3]{32})$
37. $3\sqrt{5}(2\sqrt{3} + 4\sqrt{5})$
38. $8\sqrt{6}(3\sqrt{3} - 4\sqrt{6})$
39. $-4\sqrt{12}(\sqrt{3} - 3\sqrt{8})$
40. $2\sqrt{10}(3\sqrt{2} + 5\sqrt{5})$

4.5. Rationalizing the Denominator

Occasionally, fractions are encountered whose denominators are irrational numbers. It is highly desirable to simplify these expressions so that the denominator becomes a rational number. The reasons for this preference shall become clear in the next chapter. To illustrate, the fraction $\dfrac{1}{\sqrt{2}}$ has an irrational denominator which can be made rational by using the fundamental property of fractions. That is,

$$\frac{1}{\sqrt{2}} = \frac{1}{\sqrt{2}} \cdot \frac{\sqrt{2}}{\sqrt{2}} = \frac{\sqrt{2}}{2}$$

The process of changing the denominator of a fraction from an irrational number to a rational number is called **rationalizing the denominator.** The following examples may serve to clarify how to rationalize the

Irrational Numbers 153

denominator of any fraction:

(a) $\dfrac{6}{\sqrt{3}} = \dfrac{6}{\sqrt{3}} \cdot \dfrac{\sqrt{3}}{\sqrt{3}}$

$= \dfrac{6\sqrt{3}}{3}$

$= \dfrac{6}{3} \cdot \dfrac{\sqrt{3}}{1}$

$= 2\sqrt{3}$

(b) $\dfrac{4}{\sqrt[3]{2}} = \dfrac{4}{\sqrt[3]{2}} \cdot \dfrac{\sqrt[3]{4}}{\sqrt[3]{4}}$

$= \dfrac{4 \cdot \sqrt[3]{4}}{\sqrt[3]{8}}$

$= \dfrac{4 \cdot \sqrt[3]{4}}{2}$

$= 2\sqrt[3]{4}$

(c) First method: $\dfrac{5}{\sqrt{12}} = \dfrac{5}{\sqrt{12}} \cdot \dfrac{\sqrt{12}}{\sqrt{12}}$

$= \dfrac{5 \cdot \sqrt{12}}{12}$

$= \dfrac{5\sqrt{4} \cdot \sqrt{3}}{12}$

$= \dfrac{5 \cdot 2 \cdot \sqrt{3}}{12}$

$= \dfrac{5\sqrt{3}}{6}$

Second method: $\dfrac{5}{\sqrt{12}} = \dfrac{5}{\sqrt{12}} \cdot \dfrac{\sqrt{3}}{\sqrt{3}}$

$= \dfrac{5\sqrt{3}}{\sqrt{36}}$

$= \dfrac{5\sqrt{3}}{6}$

154   Pre-Algebra

Observe in (a) and (b) that the numerator and denominator are multiplied by the smallest number necessary to produce a perfect power. Note that the first method in (c) is not the most efficient because $\sqrt{12}$ is not the smallest number which will produce a perfect square. A more direct result is found in the second method by multiplying by $\sqrt{3}$, which is the smallest number which will produce a perfect square.

Fractions such as $\dfrac{\sqrt{2}}{2}$ can also be expressed in a different manner by considering that

$$\frac{\sqrt{2}}{2} = \sqrt{2} \div 2 = \sqrt{2} \cdot \frac{1}{2} = \frac{1}{2}\sqrt{2}$$

Simply, then $\dfrac{\sqrt{2}}{2} = \dfrac{1}{2}\sqrt{2}$, and also $\dfrac{2\sqrt{3}}{5} = \dfrac{2}{5}\sqrt{3}$.

This equivalent way of stating fractions can be useful when simplifying the following kinds of expressions:

(d) $\dfrac{\sqrt{2}}{5} - \dfrac{\sqrt{2}}{6} = \dfrac{1}{5}\sqrt{2} - \dfrac{1}{6}\sqrt{2}$

$\qquad\qquad = \left(\dfrac{1}{5} - \dfrac{1}{6}\right) \cdot \sqrt{2}$ By the distributive property

$\qquad\qquad = \left(\dfrac{6}{30} - \dfrac{5}{30}\right) \cdot \sqrt{2}$

$\qquad\qquad = \dfrac{1}{30}\sqrt{2}$ or $\dfrac{\sqrt{2}}{30}$

(e) $\dfrac{\sqrt{8}}{3} - 4\sqrt{2} = \dfrac{2\sqrt{2}}{3} - 4\sqrt{2}$

$\qquad\qquad = \dfrac{2}{3}\sqrt{2} - 4\sqrt{2}$

$\qquad\qquad = \left(\dfrac{2}{3} - 4\right)\sqrt{2}$ By the distributive property

$\qquad\qquad = \left(\dfrac{2}{3} - \dfrac{12}{3}\right)\sqrt{2}$

$\qquad\qquad = \dfrac{-10}{3}\sqrt{2}$ or $\dfrac{-10\sqrt{2}}{3}$

Irrational Numbers 155

Now observe that $\left(\sqrt{\dfrac{2}{3}}\right)^2 = \dfrac{2}{3}$ and $\left(\dfrac{\sqrt{2}}{\sqrt{3}}\right)^2 = \dfrac{(\sqrt{2})^2}{(\sqrt{3})^2} = \dfrac{2}{3}$.

Since both $\sqrt{\dfrac{2}{3}}$ and $\dfrac{\sqrt{2}}{\sqrt{3}}$ are positive numbers whose squares are equal, then it must be that $\sqrt{\dfrac{2}{3}} = \dfrac{\sqrt{2}}{\sqrt{3}}$. Also, $\left(\dfrac{\sqrt[3]{-3}}{\sqrt[3]{5}}\right)^3 = \dfrac{(\sqrt[3]{-3})^3}{(\sqrt[3]{5})^3} = \dfrac{-3}{5}$

so that $\sqrt[3]{\dfrac{-3}{5}} = \dfrac{\sqrt[3]{-3}}{\sqrt[3]{5}}$. This principle holds whenever the radicand is positive and n is even, or for any rational number when n is odd. Generally then,

$$\sqrt[n]{\dfrac{a}{b}} = \dfrac{\sqrt[n]{a}}{\sqrt[n]{b}}$$

n a natural number such that $n \geq 3$
when n is odd, a and b are any two rational numbers, $b \neq 0$
when n is even, a is nonnegative and b is positive,
where both a and b are rational numbers

Finally, when fractions occur in the radicand, they should be eliminated. Consider the following examples which illustrate this kind of calculation with the aid of the above property:

(f) $\sqrt{\dfrac{2}{3}} = \sqrt{\dfrac{6}{9}}$ Since $\dfrac{2}{3} = \dfrac{6}{9}$

$= \dfrac{\sqrt{6}}{\sqrt{9}}$ $\sqrt{\dfrac{a}{b}} = \dfrac{\sqrt{a}}{\sqrt{b}}$

$= \dfrac{\sqrt{6}}{3}$ or $\dfrac{1}{3}\sqrt{6}$

(g) $\sqrt{\dfrac{3}{8}} = \sqrt{\dfrac{6}{16}}$ Since $\dfrac{3}{8} = \dfrac{6}{16}$

$= \dfrac{\sqrt{6}}{\sqrt{16}}$ $\sqrt{\dfrac{a}{b}} = \dfrac{\sqrt{a}}{\sqrt{b}}$

$= \dfrac{\sqrt{6}}{4}$ or $\dfrac{1}{4}\sqrt{6}$

(h) $\sqrt[3]{\dfrac{3}{4}} = \sqrt[3]{\dfrac{6}{8}}$

$= \dfrac{\sqrt[3]{6}}{\sqrt[3]{8}}$

$= \dfrac{\sqrt[3]{6}}{2}$ or $\dfrac{1}{2}\sqrt[3]{6}$

In these calculations, care should be taken to change the fraction into one whose denominator is the smallest perfect power.

In summation, expressions containing radicals can be added, subtracted, multiplied, and divided. In order to say that a radical has been "simplified," it must satisfy all the three following rules:

1. No perfect powers remain in the radicand.
2. The denominator of a fraction has been rationalized.
3. No fraction remains in the radicand.

4.5. Exercises

Simplify each of the following radicals by following the three rules given above:

1. $\dfrac{4}{\sqrt{2}}$
2. $\dfrac{2}{\sqrt{6}}$
3. $\dfrac{3}{\sqrt{8}}$
4. $\dfrac{4}{\sqrt{12}}$

5. $\dfrac{1}{2\sqrt{2}}$
6. $\dfrac{3}{4\sqrt{6}}$
7. $\dfrac{6}{\sqrt[3]{3}}$
8. $\sqrt{\dfrac{3}{4}}$

9. $\sqrt{\dfrac{1}{3}}$
10. $\sqrt{\dfrac{5}{8}}$
11. $\sqrt[3]{\dfrac{3}{4}}$
12. $\dfrac{1}{\sqrt[3]{2}}$

13. $\dfrac{-4}{\sqrt[3]{40}}$
14. $\dfrac{\sqrt{3}}{\sqrt{6}}$
15. $\dfrac{\sqrt[3]{2}}{\sqrt[3]{3}}$
16. $\dfrac{\sqrt{3} \cdot \sqrt{2}}{\sqrt{8}}$

Write each of the following products or powers as simplified radicals or rational numbers:

17. $\left(\dfrac{4}{\sqrt{3}}\right)^2$
18. $\left(\dfrac{-3}{\sqrt{2}}\right)^3$

19. $\left(\dfrac{15}{\sqrt{3}}\right)^2$
20. $\left(\dfrac{-4}{3\sqrt{2}}\right)^2$

21. $\left(\dfrac{-3}{2\sqrt{6}}\right)^3$
22. $\left(\dfrac{\sqrt{6}}{\sqrt{8}}\right)^3$

23. $\left(\dfrac{\sqrt{3} \cdot \sqrt{5}}{\sqrt{75}}\right)^3$
24. $\left(\dfrac{2\sqrt{3}}{\sqrt{12}}\right)^4$

Irrational Numbers   157

Combine each of the following expressions into a single, simplified fraction:

25. $\dfrac{\sqrt{5}}{4} + \dfrac{\sqrt{5}}{6}$

26. $\dfrac{3\sqrt{6}}{4} + \dfrac{\sqrt{6}}{2}$

27. $\dfrac{\sqrt{12}}{5} - \dfrac{\sqrt{3}}{6}$

28. $\dfrac{2\sqrt{8}}{7} + \dfrac{\sqrt{2}}{35}$

29. $\dfrac{\sqrt{6}}{3} - \dfrac{\sqrt{24}}{3}$

30. $\dfrac{\sqrt{18}}{6} - \dfrac{\sqrt{32}}{5}$

31. $\dfrac{\sqrt{3}}{4} - 5\sqrt{3}$

32. $6\sqrt{7} - \dfrac{\sqrt{7}}{5}$

4.6. Evaluation

It is difficult to think of a field of science or mathematics in which it is not necessary at times to evaluate certain kinds of mathematical expressions. To evaluate is to "find the value of" an expression for a given set of values. Of particular concern here are expressions which contain exponents and radicals.

One kind of mathematical expression is a formula. A formula might be defined as a mathematical principle expressed in algebraic symbols. The following are some common formulas:

(a) $\quad d = r \cdot t$
(b) $\quad s = \frac{1}{2} \cdot a \cdot t^2$
(c) $\quad c = \sqrt{a^2 + b^2}$

In (a), d represents the distance, in miles, that an object has traveled after going at a rate of r miles per hour for t hours. When $r = 60$ miles per hour (mph), and $t = 3$ hours, then $d = 60 \cdot 3 = 180$ miles.

In (b), s represents the distance, in feet, that an object travels in t seconds, when it is accelerated at a feet per second per second (usually written ft/sec²). When $a = 3$ ft/sec² and $t = 4$ seconds, then the distance $s = \frac{1}{2} \cdot 3 \cdot 4^2 = 24$ feet.

158   Pre-Algebra

In (c), *a*, *b*, and *c* represent the three sides of a triangle as shown below:

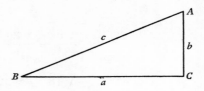

When angle C is 90° (a right angle), the triangle is called a right triangle. Then c is called the hypotenuse, and a and b are called the legs of the right triangle. This statement, called the Pythagorean theorem, states that the hypotenuse of a right triangle is equal to the square root of the sum of the squares of the lengths of the two legs. This theorem is over 2,000 years old. If the two legs of a right triangle are 4 and 6 inches, then the length of the hypotenuse is

$$\begin{aligned} c &= \sqrt{4^2 + 6^2} \\ &= \sqrt{16 + 36} \\ &= \sqrt{52} \\ &= \sqrt{4} \cdot \sqrt{13} \\ &= 2\sqrt{13} \text{ inches} \end{aligned}$$

Note that both the 4 and the 6 are squared and then added before finding the square root. In general, the square root of the sum of two numbers is not equal to the sum of their square roots, which is discussed in Section 4.4, p. 148.

These examples are but three of an endless number of applications where exponents and radicals are used.

In algebra, expressions which are called **polynomials** frequently require evaluation. The following expressions are examples of polynomials:

(d) $P(x) = 3x^2 + x + 7$
(e) $P(x) = x^4 + 3x^2 - 8x - 5$
(f) $P(x,y) = x^2y + 3xy - 2xy^2$

In the expressions above, x and y are called **variables** because they can be replaced by more than one numeral and therefore their value varies. Symbols which only have one value, such as -3, $\sqrt{5}$, and $\frac{4}{7}$, are called **constants.** A polynomial is any sum of algebraic expressions each of which is written as a product of numbers and nonnegative integral powers of the variables. Examples (d) and (e) are polynomials which are stated in terms of a variable x, named $P(x)$, and read as "P of x." The notation, $P(x)$, simply identifies a polynomial whose variable is x and it should not be mistaken for the meaningless product of P and x. In (d) the constants are 2, 3, and 7, while in (e), the constants are 2, 3, 4, -8, and -5. In order to evaluate $P(x)$ for a specific value, x is replaced by that specific value wherever x occurs in the polynomial. If the polynomial is evaluated with $x = 3$, then its value is identified by the notation, $P(3)$. The notation $P(-2)$ represents the value of the polynomial at $x = -2$. In order to find a specific value of example (f), it is necessary to have specific values for both x and y. If x is replaced by 3 and y is replaced by -2, then $P(x,y)$, which is read as "P of x,y," becomes $P(3,-2)$. The following examples demonstrate the evaluation procedure:

(g) $P(x) = x^2 + 4x + 5$ where x is any rational number
When $x = 3$,
$$P(3) = 3^2 + 4(3) + 5$$
$$= 9 + 12 + 5$$
$$= 26$$

(h) $P(x) = 2x^3 - 3x^2 - 4x - 1$
When $x = 2$,
$$P(2) = 2(2)^3 - 3(2)^2 - 4(2) - 1$$
$$= 2 \cdot 8 - 3 \cdot 4 - 8 - 1$$
$$= 16 - 12 - 8 - 1$$
$$= -5$$

(i) $P(x,y) = 3x^2y - 4xy - 5xy^2$
When $x = 2$ and $y = -3$,
$$P(2,-3) = 3(2)^2(-3) - 4(2)(-3) - 5(2)(-3)^2$$
$$= -36 - (-24) - 90$$
$$= -102$$

Other kinds of expressions in algebra, called **algebraic expressions,** also may require evaluation. Any expression obtained by a finite number of arithmetic operations, or the extraction of roots, is called an algebraic expression. Many algebraic expressions contain more than one variable. An algebraic expression in two variables, x and y, is denoted by $R(x,y)$. If the expression is evaluated at $x = 3$ and $y = 4$, it is identified by $R(3,4)$, taking care that the x value is expressed first and the y value second. The following examples will illustrate this type of evaluation:

(j) $R(x,y) = \dfrac{x^2 - y^2}{\sqrt{x^2 + y^2}}$ where x and y are any two rational numbers, not both zero

When $x = 3$ and $y = 4$,

$$R(3,4) = \dfrac{3^2 - 4^2}{\sqrt{3^2 + 4^2}}$$

$$= \dfrac{9 - 16}{\sqrt{9 + 16}}$$

$$= \dfrac{-7}{\sqrt{25}}$$

$$R(3,4) = \dfrac{-7}{5}$$

(k) $R(x,y) = \dfrac{x - y^3}{2x} + \dfrac{x^3 - y}{3y}$ where x and y are any two nonzero rational numbers

When $x = 3$ and $y = -2$,

$$R(3,-2) = \dfrac{3 - (-2)^3}{2 \cdot 3} + \dfrac{3^3 - (-2)}{3(-2)}$$

$$= \dfrac{3 - (-8)}{6} + \dfrac{27 - (-2)}{-6}$$

$$= \dfrac{11}{6} + \dfrac{29}{-6}$$

$$= \dfrac{11}{6} + \dfrac{-29}{6}$$

$$= \dfrac{11 + (-29)}{6}$$

$$= \dfrac{-18}{6}$$

$$R(3,-2) = -3$$

4.6. Exercises

Evaluate each of the following expressions for the values given. State your answer as a simplified radical or as a reduced rational number.

Evaluate $c = \sqrt{a^2 + b^2}$ for the values of a and b given below:

1. $a = 6, b = 8$
2. $a = 2, b = 4$
3. $a = 3, b = 9$

Evaluate $K = \sqrt{s(s-a)(s-b)(s-c)}$ where $s = \frac{1}{2}(a+b+c)$ for the values of a, b, and c given below:

4. $a = 3, b = 4, c = 5$
5. $a = 5, b = 6, c = 7$
6. $a = 2, b = 3, c = 4$

Evaluate $P(x) = x^2 + 3x + 5$ for the values of x given below:

7. $P(4)$
8. $P(-3)$

Evaluate $P(x) = 2x^2 - 4x - 3$ for the values of x given below:

9. $P(3)$
10. $P(-3)$

Evaluate $P(x) = 2x^3 - 3x^2 - 4x - 1$ for the values of x given below:

11. $P(3)$
12. $P(-2)$

Evaluate $R(x) = \dfrac{(3-x)^3}{x^2} + \dfrac{(2x-3)^2}{x^3}$ for the values of x given below:

13. $R(4)$
14. $R(-1)$

Evaluate $R(x,y) = \dfrac{x^2 - y^2}{\sqrt{x^2 + y^2}}$ where x and y are any two rational numbers, not both zero. The value of x and y are given below:

15. $R(12,5)$
16. $R(2,4)$

Evaluate $R(x,y) = \dfrac{x - y^3}{2x} + \dfrac{x^3 - y}{3y}$ where x and y are any two nonzero rational numbers, whose values are given below:

17. $R(4,2)$
18. $R(-2,-1)$

Evaluate $R(x,y) = \dfrac{(x-3y)^2}{\sqrt{5x-y}} + \dfrac{(2x-y)^2}{\sqrt{x+2y}}$ where x and y are any two rational numbers, not both zero, whose values are given below:

19. $R(3,3)$
20. $R(2,1)$

162 Pre-Algebra

4.7. Summary

This chapter has been devoted to a brief introduction of exponents and radicals. Algebraic expressions were also presented here in order to learn how to evaluate them for specific values of their variables. These kinds of numerial calculations shall prove to be invaluable when studying algebra and other courses in mathematics.

It was necessary to introduce radical notation when exact equal rational factors of a given number could not be found. These new quantities which cannot be expressed as rational numbers, are called irrational numbers. Although irrational numbers do not have exact rational values, the usual arithmetic operations can be performed with them. A new notation shall be developed in Chapter 5 in order to express the approximate value of irrational numbers as well as the exact value of rational numbers.

4.7. Review Exercises

Write each of the following products in exponential notation. Write each quantity with a single exponent and no more than a single sign:

1. $4 \cdot 4 \cdot 4$
2. $(-3)(-3)(-3)(-3)$
3. $-5 \cdot 5 \cdot 5 \cdot 5$
4. $\dfrac{-3}{5} \cdot \dfrac{-3}{5} \cdot \dfrac{-3}{5}$
5. $7^2 \cdot 7^6$
6. $6^3 \cdot 6^4 \cdot 6^2$
7. $(8^3)^4$
8. $8^3 \cdot 8^4$
9. $\dfrac{-5^3}{-5^7}$
10. $\left(\dfrac{2}{3^2}\right)^3$
11. $\dfrac{(-4)^5}{(-4)^2}$
12. $(4^2 \cdot 7)^3$
13. $(3^3 \cdot 4^3 \cdot 5)^4$
14. $(4^3)^2 \cdot (4^2)^4$
15. $3(-4) \cdot 3 \cdot 3 \cdot 5 \cdot 5 \cdot (-4) \cdot 5 \cdot 5$
16. $\dfrac{-4^3}{4}$

Irrational Numbers 163

Simplify each of the following radicals by the rules given in Section 4.5, p. 156:

17. $\sqrt{36}$

18. $-\sqrt{64}$

19. $-\sqrt[3]{64}$

20. $\sqrt{75}$

21. $\sqrt{20}$

22. $-3\sqrt{40}$

23. $4\sqrt[3]{40}$

24. $-2\sqrt[3]{54}$

25. $\dfrac{3}{\sqrt{3}}$

26. $\dfrac{4}{\sqrt{8}}$

27. $\dfrac{6}{\sqrt{3}}$

28. $\dfrac{2}{3\sqrt{6}}$

29. $\sqrt{\dfrac{2}{3}}$

30. $\sqrt[3]{\dfrac{2}{3}}$

31. $\dfrac{\sqrt{2}}{\sqrt{6}}$

32. $\sqrt{\dfrac{5}{18}}$

Write each of the following sums or differences as a product wherever possible. Simplify any radical in your final answer:

33. $\sqrt{3} + 3\sqrt{3} + 5\sqrt{3}$

34. $\sqrt{12} + \sqrt{3} - \sqrt{27}$

35. $\sqrt[3]{16} - 4\sqrt[3]{2} + \sqrt[3]{54}$

36. $\sqrt{18} + \sqrt{8} + 3\sqrt{2}$

37. $\dfrac{\sqrt{5}}{3} - \dfrac{\sqrt{5}}{2}$

38. $\dfrac{2\sqrt{3}}{3} - \dfrac{\sqrt{3}}{6}$

39. $\dfrac{\sqrt{24}}{8} + \dfrac{\sqrt{54}}{9}$

40. $\dfrac{3}{\sqrt{8}} + \sqrt{\dfrac{1}{2}}$

Simplify each of the following products by first using the distributive property and then reducing and combining radicals wherever possible:

41. $\dfrac{1}{12}(8\sqrt{2} + 9\sqrt{8})$

43. $2\sqrt{7}(2\sqrt{7} - 3\sqrt{14})$

42. $\sqrt{5}(\sqrt{20} + 3\sqrt{10})$

44. $-3\sqrt{6}(2\sqrt{72} - 5\sqrt{10})$

Write each of the following products or powers as simplified radicals or integers:

45. $\sqrt{11} \cdot \sqrt{11}$
46. $\sqrt{43} \cdot \sqrt{43}$
47. $\sqrt{3} \cdot \sqrt{15}$
48. $(3\sqrt{2})(-5\sqrt{12})$
49. $(2\sqrt{3})^3$
50. $(\sqrt{6})^4$

51. $(3\sqrt{6})^2$
52. $\sqrt{6}\sqrt{3}\sqrt{18}$
53. $(-2\sqrt{5})^3$
54. $(4\sqrt{8})(3\sqrt{5})$
55. $(5\sqrt{2})^2(-3\sqrt{6})$
56. $(\sqrt[3]{4})^2$

Evaluate each of the following expressions for the values given. State your answer as a simplified radical or as a reduced rational number:

Given: $y = \sqrt{26 - 2x^2}$. Find y for each of the values of x given below:

57. $x = -2, y =$

58. $x = 3, y =$

Evaluate $P(x) = 3x^2 - 2x - 1$ for the values of x given below:

59. $P(2)$

60. $P(-3)$

Evaluate $P(x) = -2x^3 + x^2 + 6x - 3$ for the values of x given below:

61. $P(-1)$

62. $P(2)$

Evaluate $P(x, y) = \dfrac{(x - 3y)^2}{\sqrt{3x + y}}$ for the values of x and y given below:

63. $P(2, -1)$

64. $P(3, 3)$

5 Real Numbers

5.1. Nonpositive Exponents

Irrational numbers, as discussed in Chapter 4, do not have exact fractional equivalents. In certain kinds of mathematical calculations, it is inconvenient to work with fractions. For these reasons, and others, it is desirable to invent a new notation which could be used instead of the fractional form. Before discussing this new notation, it will be necessary to introduce a new concept, that of the negative exponent. A negative exponent is defined in terms of a positive exponent as follows:

> A number raised to any negative power is defined as the reciprocal of that number raised to the opposite of that power.

Symbolically then,

$$a^{-1} = \frac{1}{a} \qquad a^{-2} = \frac{1}{a^2} \qquad a^{-3} = \frac{1}{a^3}$$

Pre-Algebra

and in general

$$a^{-n} = \frac{1}{a^n}$$ where a is any nonzero rational number and n is any nonzero integer

To illustrate, $10^{-1} = \frac{1}{10}$ or $\frac{1}{10^1}$, $6^{-2} = \frac{1}{6^2}$, $-3^{-4} = \frac{1}{-3^4}$, and $(\frac{2}{3})^{-1} = \frac{1}{\frac{2}{3}} = \frac{3}{2}$.

Observe that 10^{-1} is $\frac{1}{10}$, the reciprocal of 10. Also note that $(\frac{2}{3})^{-1}$ is $\frac{3}{2}$, the reciprocal of $\frac{2}{3}$. It should seem reasonable then that

$$a^{-1}$$ is the reciprocal of a for all rational numbers, $a \neq 0$

It is desirable that operations with negative exponents have the same properties as positive exponents. The remainder of this section is concerned with the extension of these properties.

Consider the product of numbers which are raised to negative powers. Since $10^{-1} = \frac{1}{10}$, then $10^{-1} \cdot 10^{-1} = \frac{1}{10} \cdot \frac{1}{10} = \frac{1}{10^2}$. Now $\frac{1}{10^2} = 10^{-2}$, by definition, so that $10^{-1} \cdot 10^{-1} = 10^{-2}$. Now consider the product $6^{-2} \cdot 6^{-3}$. Since $6^{-2} = \frac{1}{6^2}$ and $6^{-3} = \frac{1}{6^3}$, then $6^{-2} \cdot 6^{-3} = \frac{1}{6^2} \cdot \frac{1}{6^3} = \frac{1}{6^5}$ and $\frac{1}{6^5} = 6^{-5}$. Therefore, $6^{-2} \cdot 6^{-3} = 6^{-5}$. In Section 4.1, p. 131, it was developed that $a^m \cdot a^n = a^{m+n}$, where m and n are natural numbers. It should now seem reasonable that generally,

$$a^m \cdot a^n = a^{m+n}$$ where a is any nonzero rational number, m and n any two nonzero integers

To illustrate this, consider the following examples:

(a) $\dfrac{4^3}{4^5} = \dfrac{1}{4^2}$ $\qquad\qquad \dfrac{a^m}{a^n} = \dfrac{1}{a^{n-m}}, m < n$

But also,

$\dfrac{4^3}{4^5} = \dfrac{4^3}{1} \cdot \dfrac{1}{4^5}$

$\qquad = 4^3 \cdot 4^{-5}$

$\qquad = 4^{-2} \qquad\qquad a^m \cdot a^n = a^{m+n}$

(b) $\dfrac{1}{7^3} \cdot \dfrac{1}{7^5} = \dfrac{1}{7^8}$ or 7^{-8}

But also,

$\dfrac{1}{7^3} \cdot \dfrac{1}{7^5} = 7^{-3} \cdot 7^{-5}$

$\qquad\qquad = 7^{-8}$

(c) $\dfrac{4}{100} = \dfrac{4}{10^2}$

$\qquad = 4 \cdot \dfrac{1}{10^2}$

$\qquad = 4 \cdot 10^{-2}$

The above examples illustrate that numbers raised to positive exponents, which occur in the denominator of a fraction, can be moved to the numerator and expressed with negative exponents. As a matter of fact, a number which is raised to a power can be moved to the opposite term of a fraction by changing the sign of its exponent to the opposite. The following examples illustrate this:

(d) $\dfrac{5}{7^{-2}} = \dfrac{5}{\dfrac{1}{7^2}}$

$\qquad = 5 \cdot 7^2$

Also note that $5 \cdot 7^2 = \dfrac{5}{7^{-2}}$

(e) $4^3 \cdot 6^{-4} = 4^3 \cdot \dfrac{1}{6^4}$

$\qquad\qquad = \dfrac{4^3}{6^4}$

So that generally,

$a^{-n} = \dfrac{1}{a^n}$ or $a^n = \dfrac{1}{a^{-n}}$ where *a* is any nonzero number, and *n* any nonzero integer

The property $\frac{a^m}{a^n} = a^{m-n}$ holds when $m > n$. It is desirable to extend this result to the case where $m = n$. By this property and the assumption that m can equal n, $\frac{7^2}{7^2} = 7^{2-2} = 7^0$, read as "7 to the zero power." Since $\frac{7^2}{7^2} = 1$, it should be clear that $7^0 = 1$. Again, $\frac{10^4}{10^4} = 10^{4-4} = 10^0$, but also $\frac{10^4}{10^4} = 1$, so that $10^0 = 1$. Generally then,

$$a^0 = 1 \quad \text{where } a \text{ is any nonzero number}$$

Restated verbally, any number, with the exception of zero, can be raised to the zero power, and the result is one.

Using these results, a more general statement can be made regarding negative exponents. That is,

$$a^{-n} = \frac{1}{a^n} \quad \text{or} \quad a^n = \frac{1}{a^{-n}} \quad \text{where } a \text{ is any nonzero number, and } n \text{ is any integer}$$

Therefore, $2^3 = 2^{-(-3)} = \frac{1}{2^{-3}}$ and $\frac{1}{10^{-6}} = 10^{-(-6)} = 10^6$.

Using the negative exponent, the division of numbers in exponential notation can now be expressed by the single property,

$$\frac{a^m}{a^n} = a^{m-n} \quad \text{where } a \text{ is any nonzero number, } m \text{ and } n \text{ are any two integers}$$

The following properties shall be assumed to be valid for exponents which are integers (their proofs shall be presented in a more advanced mathematics course):

$$a^m \cdot a^n = a^{m+n}$$
$$(a \cdot b)^n = a^n \cdot b^n$$
$$(a^m)^n = a^{m \cdot n}$$
$$\left(\frac{a}{b}\right)^n = \frac{a^n}{b^n}$$

where a and b are any two nonzero numbers, m and n are any two integers

In summary then the meaning of the zero exponent and the negative exponent have been defined in this section. With this extension of the meaning of exponents, any integer can now be used as an exponent. The properties of exponents, developed in Chapter 4 for positive exponents, are now valid for exponents which are integers.

These properties now permit numbers which are fractions or quotients to be written as products, and if so desired, products to be written as fractions. This ability becomes a useful tool in mathematics.

5.1. Exercises

Express each of the following with a single positive exponent:

1. 4^{-3}

2. $\dfrac{1}{3^{-2}}$

3. 7^{-1}

4. $\left(\dfrac{4}{7}\right)^{-1}$

5. $\left(\dfrac{3}{5}\right)^{-3}$

6. -5^{-2}

7. $\dfrac{1}{-8^{-4}}$

8. $7^3 \cdot 7^{-5}$

9. $\left(\dfrac{1}{4}\right)^{-1}$

10. $2^{-6} + 2^{-6}$

11. $3^{-2} + 3^{-2} + 3^{-2}$

12. $(3^4 \cdot 4^{-2})^{-1}$

13. 10^{-4}

14. $9^{-2} \cdot 9^{-3}$

15. $10^{-3} \cdot 10^5$

16. $\left(\dfrac{-5}{7}\right) \cdot \left(\dfrac{-5}{7}\right)^{-3}$

Simplify each of the following expressions so that each numeral is represented only once in exponential notation and your answer is not in fractional form $\left(\text{Example: } \dfrac{4^3}{4^5 \cdot 7^2} = 4^3 \cdot 4^{-5} \cdot 7^{-2} = 4^{-2} \cdot 7^{-2}\right)$:

17. $\dfrac{10^3}{10^5}$

24. $\dfrac{4^{-3} \cdot 5^2}{8^0 \cdot 4^7}$

18. $\dfrac{(4^2 + 5)^0}{7^3}$

25. $(2^3)^{-4}$

19. $\dfrac{3^4}{3^{-2} \cdot 5^{-3}}$

26. $\dfrac{4^5}{\sqrt{(7^2 + 3^2)^0}}$

20. $\dfrac{7^0}{5^3 \cdot 3^5}$

27. $(7 \cdot 4^2)^{-3}$

21. $\dfrac{10^2}{10^{-4}}$

28. $\left(\dfrac{6^2}{7^3}\right)^{-4}$

22. $\dfrac{2^{-7}}{8^0 + 3^0}$

29. $(3^0 \cdot 8^4 \cdot 5^{-3})^3$

23. $\dfrac{10^{-4}}{10^{-7}}$

30. $\left(\dfrac{5^{-3}}{10^0}\right)^{-2}$

5.2. Decimal Fractions

Expanded notation was used in Chapter 1 to write natural numbers such as 4,328 as follows: 4,328 = 4,000 + 300 + 20 + 8. Using exponents, it can now be written as $4{,}328 = 4 \cdot 10^3 + 3 \cdot 10^2 + 2 \cdot 10^1 + 8 \cdot 10^0$. Observe that the exponent of the base 10 of each successive digit, proceeding to the right, is reduced by a power of 1. It seems natural to permit the exponent of the base to continue to decrease into the negative powers of 10. By doing this, numbers having a numerical value of less than 1 can be written as the sum of multiples of negative powers of 10. To illustrate:

(a) $\dfrac{328}{1,000} = \dfrac{1}{1,000} \cdot (328)$ $\dfrac{a}{b} = \dfrac{1}{b} \cdot a$

$= \dfrac{1}{1,000} \cdot (300 + 20 + 8)$

$= \dfrac{300}{1,000} + \dfrac{20}{1,000} + \dfrac{8}{1,000}$ $a(b + c + d) = a \cdot b +$
$ + a \cdot c + a \cdot d$

$= \dfrac{3}{10} + \dfrac{2}{100} + \dfrac{8}{1,000}$

$= 3 \cdot \dfrac{1}{10} + 2 \cdot \dfrac{1}{10^2} + 8 \cdot \dfrac{1}{10^3}$

$= 3 \cdot 10^{-1} + 2 \cdot 10^{-2} + 8 \cdot 10^{-3}$

Note that the sum can be written in positional notation as 328, but it would naturally be mistaken for the numeral which represents three hundred twenty-eight. To make this distinction, a dot, or **decimal point** is inserted immediately to the left of the 3 and the number is now written as 0.328. A zero is generally placed to the left of the decimal point so that the decimal point will not be overlooked. A fraction such as $\dfrac{328}{1,000}$ is called a **common fraction,** whereas its equivalent, 0.328, is called a **decimal fraction.**

Some examples will help to illustrate this concept.

(b) $\dfrac{4,032}{10,000} = \dfrac{4,000}{10,000} + \dfrac{30}{10,000} + \dfrac{2}{10,000}$

$= 4 \cdot \dfrac{1}{10} + 3 \cdot \dfrac{1}{1,000} + 2 \cdot \dfrac{1}{10,000}$

$= 4 \cdot 10^{-1} + 0 \cdot 10^{-2} + 3 \cdot 10^{-3} + 2 \cdot 10^{-4}$

$= 0.4032$

174 Pre-Algebra

(c) $\dfrac{976}{10{,}000} = \dfrac{900}{10{,}000} + \dfrac{70}{10{,}000} + \dfrac{6}{10{,}000}$

$= 9 \cdot \dfrac{1}{100} + 7 \cdot \dfrac{1}{1{,}000} + 6 \cdot \dfrac{1}{10{,}000}$

$= 0 \cdot 10^{-1} + 9 \cdot 10^{-2} + 7 \cdot 10^{-3} + 6 \cdot 10^{-4}$

$= 0.0976$

By observing the patterns displayed in these examples and with practice, it will soon be possible to write common fractions whose denominators are powers of 10 directly into decimal fractions. To demonstrate this,

(d) $\dfrac{3}{10} = 0.3$ (e) $\dfrac{9}{100} = 0.09$ (f) $\dfrac{7}{1{,}000} = 0.007$

Observe that the first digit to the right of the decimal point represents multiples of tenths (one-tenth, two-tenths, etc.), the second digit to the right of the decimal point represents multiples of hundredths, etc. The diagram below shows the place value of some of the digits to the right of the decimal point:

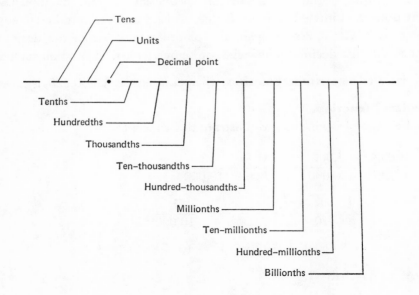

Real Numbers 175

By memorizing these place values, numbers written as decimal fractions can be quickly converted into common fractions when desired, as shown in the following examples:

(g) 0.32, read as "thirty-two hundredths," is equivalent to $\dfrac{32}{100}$

(h) 0.407, read as "four hundred seven thousandths," is equivalent to $\dfrac{407}{1,000}$

(i) 0.7392, read as "seven thousand three hundred ninety-two ten-thousandths," is equivalent to $\dfrac{7,392}{10,000}$

(j) 0.00305, read as "three hundred five hundred-thousandths," is equivalent to $\dfrac{305}{100,000}$

In this way, common fractions whose denominators are a power of 10 can be changed into decimal fractions, and decimal fractions can just as easily be changed into common fractions whose denominators are a power of 10.

Common fractions whose denominators are not a power of 10 can sometimes be changed into those which are by using the fundamental property of fractions.

The following examples will illustrate this:

(k) $\dfrac{1}{5} = \dfrac{1}{5} \cdot \dfrac{2}{2}$

$= \dfrac{2}{10}$

$= 0.2$

(l) $\dfrac{1}{4} = \dfrac{1}{4} \cdot \dfrac{25}{25}$

$= \dfrac{25}{100}$

$= 0.25$

(m) $\dfrac{3}{125} = \dfrac{3}{5^3} \cdot \dfrac{2^3}{2^3}$

$= \dfrac{24}{10^3}$

$= 0.024$

Since the factors of 10 are 2 and 5, the factors of any power of 10 will be a power of the factors of 2 and 5. So that $10^3 = 2^3 \cdot 5^3$ and in general, $10^n = 2^n \cdot 5^n$, where n is any natural number. Therefore, in order to be able to change a denominator into a power of 10, it can only contain factors of 2 and/or factors of 5. Observe that the factors of the denominators of (k), (l), and (m), respectively, are 5, 2^2, and 5^3. With the proper choice of multipliers, these denominators become $2 \cdot 5$, $2^2 \cdot 5^2$, and $2^3 \cdot 5^3$, each powers of 10. As a final illustration, $40 = 2^3 \cdot 5$, and by multiplying it by the factor 5^2, it then becomes $2^3 \cdot 5^3$ which is 10^3.

Therefore,

(n) $\dfrac{7}{40} = \dfrac{7}{2^3 \cdot 5}$

$= \dfrac{7}{2^3 \cdot 5} \cdot \dfrac{5^2}{5^2}$

$= \dfrac{175}{2^3 \cdot 5^3}$

$= \dfrac{175}{1{,}000}$

$= 0.175$

Examples (g) through (n) produce decimal representations which have a finite number of nonzero digits to the right of the decimal point. These are called **terminating** decimals. If a decimal representation does not terminate, it is called a **nonterminating** decimal. Those common fractions whose denominators cannot be changed into powers of 10 are decimals having patterns of repetititon which are nonterminating. These fractions are called **repeating, nonterminating** decimals. The simplest way to find these decimal representations is to treat the common fraction as the quotient of two integers and then use the division algorithm.

Examples of these kinds of fractions follow:

(o) $\dfrac{1}{3}$

$$\begin{array}{r} 0.3\,3\,3 \\ 3\overline{)1.0\,0\,0} \\ 9 \\ \hline 1\,0 \\ 9 \\ \hline 1\,0 \\ 9 \\ \hline 1 \end{array}$$

A bar is placed over the 3 to indicate that this digit repeats itself indefinitely. Therefore, $\dfrac{1}{3} = 0.\overline{3}$.

(p) $\dfrac{4}{33}$

$$\begin{array}{r} 0.1\,2\,1\,2 \\ 33\overline{)4.0\,0\,0\,0} \\ 3\,3 \\ \hline 7\,0 \\ 6\,6 \\ \hline 4\,0 \\ 3\,3 \\ \hline 7 \end{array}$$

Here, the 12 repeats itself indefinitely, so the bar is placed over the two digits. Therefore, $\dfrac{4}{33} = 0.\overline{12}$.

(q) $\dfrac{1}{12}$

$$\begin{array}{r} 0.0\,8\,3\,3 \\ 12\overline{)1.0\,0\,0\,0\,0} \\ 9\,6 \\ \hline 4\,0 \\ 3\,6 \\ \hline 4\,0 \\ 3\,6 \\ \hline 4 \end{array}$$

In this example, the 3 is the only digit which repeats itself indefinitely. Therefore, $\dfrac{1}{12} = 0.08\overline{3}$.

178   Pre-Algebra

In summarizing these results, any fraction can be converted into decimal notation by the division algorithm. Those common fractions having denominators which are powers of 10, or factors of powers of 10, have an exact, terminated, decimal representation. Common fractions having denominators which are not powers of 10, nor factors of powers of 10, will have a repeating, nonterminating, decimal representation.

Restating the above:

> Any decimal fraction which terminates, or which repeats itself and does not terminate, can be expressed as a common fraction.

Many details of decimal notation have been intentionally omitted in this section. It is not the intent of this text to cover all aspects of decimal notation. Only those concepts which will prove to be useful in algebra were included here.

5.2. Exercises

Write each of the following fractions in expanded notation using negative exponents. Then write each of these in decimal notation:

1. $\dfrac{49}{100}$
2. $\dfrac{423}{1,000}$
3. $\dfrac{607}{1,000}$
4. $\dfrac{281}{10,000}$
5. $\dfrac{3}{10,000}$
6. $\dfrac{1,003}{10,000}$
7. $\dfrac{14}{100,000}$
8. $\dfrac{7,084}{1,000,000}$

Write each of the following decimals as a common fraction whose denominator is a power of 10:

9. 0.07
10. 0.403
11. 0.0437
12. 0.0031
13. 0.0006
14. 0.08351
15. 0.00732
16. 0.000481

Convert each of the following common fractions into decimal fractions:

17. $\dfrac{1}{8}$
18. $\dfrac{1}{40}$
19. $\dfrac{3}{50}$
20. $\dfrac{3}{16}$
21. $\dfrac{17}{200}$
22. $\dfrac{7}{80}$
23. $\dfrac{13}{20}$
24. $\dfrac{23}{400}$

Using the division algorithm, write each of the following common fractions as repeating, nonterminating decimals:

25. $\dfrac{1}{9}$ 27. $\dfrac{3}{11}$ 29. $\dfrac{7}{24}$ 31. $\dfrac{5}{21}$

26. $\dfrac{1}{6}$ 28. $\dfrac{8}{13}$ 30. $\dfrac{1}{7}$ 32. $\dfrac{6}{17}$

5.3. Operations with Decimals

The usual arithmetic operations can be performed on numbers which are written in decimal notation. Generally, it is far easier to perform these operations with the numbers in decimal notation than to convert them into their fractional equivalents. A multiplication algorithm for decimals can be developed here by first observing a simple pattern.

(a) $0.3 = \dfrac{3}{10} = 3 \cdot 10^{-1}$

$0.37 = \dfrac{37}{100} = \dfrac{37}{10^2} = 37 \cdot 10^{-2}$

$0.372 = \dfrac{372}{1{,}000} = \dfrac{372}{10^3} = 372 \cdot 10^{-3}$

$0.3725 = \dfrac{3{,}725}{10{,}000} = \dfrac{3{,}725}{10^4} = 3{,}725 \cdot 10^{-4}$

Please note that for each place the decimal point is moved to the right, the resulting number can be multiplied by 10^{-1} and the product is equivalent to the original number. So in (a), the decimal point was moved three places to the right and then multiplied by 10^{-3} to make the product $372 \cdot 10^{-3}$ equal to 0.372.

In this manner, the decimal point of any number can be moved to the right to make it a natural number. Using this idea, any two decimals can be multiplied by first writing them as natural numbers times an appropriate power of 10.

180   Pre-Algebra

For example:

(b) $(2.6)(37.1) = (26 \cdot 10^{-1})(371 \cdot 10^{-1})$
$= (26 \cdot 371)(10^{-1} \cdot 10^{-1})$
$= 9{,}646 \cdot 10^{-2}$

(c) $(0.0034)(8.3) = (34 \cdot 10^{-4})(83 \cdot 10^{-1})$
$= (34 \cdot 83)(10^{-4} \cdot 10^{-1})$
$= 2{,}822 \cdot 10^{-5}$

Now observe the following pattern:

(d) $3 \cdot 10^{-1} = \dfrac{3}{10} = 0.3$

$3 \cdot 10^{-2} = \dfrac{3}{10^2} = \dfrac{3}{100} = 0.03$

$3 \cdot 10^{-3} = \dfrac{3}{10^3} = \dfrac{3}{1{,}000} = 0.003$

In (d), multiplication by 10 to a negative power moves the decimal point to the left the same number of places as the numerical value of the negative power. Specifically, multiplication by 10^{-2} moves the decimal point two places to the left. Multiplication by 10^{-5} moves the decimal point five places to the left. Referring to (b) again, $9{,}646 \cdot 10^{-2} = 96.46$, since multiplication by 10^{-2} moves the decimal point two places to the left. In (c), $2{,}822 \cdot 10^{-5} = 0.02822$, and here, a 0 is included as a place holder so that the decimal point can be moved five places to the left.

A final example:

(e) $(0.043)(0.32) = (43 \cdot 10^{-3})(32 \cdot 10^{-2})$
$= (43 \cdot 32)(10^{-3} \cdot 10^{-2})$
$= 1{,}376 \cdot 10^{-5}$
$= 0.01376$

This procedure is generally simplified by counting the total number of places that the decimal point is to be moved to make both factors whole numbers, then moving the decimal point of the product that many places to the left. The reader was probably given this rule many times in previous courses and perhaps wondered why it always "worked." Now that the justification is provided, perhaps the method is more meaningful and therefore better understood.

Division of decimals can also be performed in much the same way as multiplication. The following examples illustrate this procedure:

(f) $\dfrac{3.7}{0.09} = \dfrac{37 \cdot 10^{-1}}{9 \cdot 10^{-2}}$

$= \dfrac{37 \cdot 10^{-1} \cdot 10^{2}}{9}$

$= \dfrac{37 \cdot 10}{9}$

$= \dfrac{370}{9}$

(g) $\dfrac{0.007}{8.3} = \dfrac{7 \cdot 10^{-3}}{83 \cdot 10^{-1}}$

$= \dfrac{7 \cdot 10^{-3} \cdot 10^{1}}{83}$

$= \dfrac{7 \cdot 10^{-2}}{83}$

$= \dfrac{7}{83 \cdot 10^{2}}$

$= \dfrac{7}{8,300}$

In (f), observe that the decimal point was moved two places to the right in both the numerator and denominator. In (g), it was necessary to move the decimal point in the numerator and denominator three places to the right, to make both natural numbers. This can be generalized by the following rule: if two decimals are to be divided, move the decimal point in the numerator and denominator the same number of places so that both are natural numbers. The resulting expression can then be treated as a common fraction or as a division problem involving natural numbers. If left as a common fraction, it should be reduced. If it is to be treated as a division problem, then use the division algorithm to find the quotient as a decimal. In this way, the answer can be expressed as a common fraction, a mixed number, or as a decimal. The nature of the problem will generally determine which form is most desirable.

The algorithm for the addition of decimals can be developed in much the same way. Consider the following sums:

(h) $0.134 + 0.382 = 134 \cdot 10^{-3} + 382 \cdot 10^{-3}$
$= (134 + 382) \cdot 10^{-3}$ Distributive property
$= 516 \cdot 10^{-3}$
$= 0.516$

(i) $0.0712 + 0.44$
$= 712 \cdot 10^{-4} + 44 \cdot 10^{-2}$
$= 712 \cdot 10^{-4} + 44 \cdot (10^2 \cdot 10^{-2}) \cdot 10^{-2}$ $10^2 \cdot 10^{-2} = 1$
$= 712 \cdot 10^{-4} + (44 \cdot 10^2)(10^{-2} \cdot 10^{-2})$ Associative property
$= 712 \cdot 10^{-4} + 4{,}400 \cdot 10^{-4}$
$= (712 + 4{,}400) \cdot 10^{-4}$
$= 5{,}112 \cdot 10^{-4}$
$= 0.5112$

(j) $4.38 + 0.84 = 438 \cdot 10^{-2} + 84 \cdot 10^{-2}$
$= (438 + 84) \cdot 10^{-2}$
$= 522 \cdot 10^{-2}$
$= 5.22$

By writing the sums in (h), (i), and (j) vertically, with the decimal points directly over each other, the numbers can be added with the addition algorithm finding the same results as above.

(h)   0.134   (i)   0.0712   (j)   4.38
       0.382   0.4400   0.84
       $\overline{0.516}$ $\overline{0.5112}$ $\overline{5.22}$

Subtraction with decimals can be performed with the subtraction algorithm in the same manner as with natural numbers, providing that the decimal point is lined up when writing the problem.

The following examples illustrate this last statement:

(k) $0.314 - 0.1683$

$$\begin{array}{r} 0.3140 \\ -0.1683 \\ \hline 0.1457 \end{array}$$

(l) $4.003 - 1.437$

$$\begin{array}{r} 4.003 \\ -1.437 \\ \hline 2.566 \end{array}$$

Both addition and subtraction of decimals can be performed horizontally by adding or subtracting like place values. However, it is more difficult and generally not recommended.

In this section, the arithmetic operations with decimals were developed and related to arithmetic operations with natural numbers. In each operation, both positive and negative exponents were found to be useful to help relate the algorithms to both kinds of numbers.

5.3. Exercises

Write each of the following decimals as the product of a natural number and a power of 10:

1. 0.57
2. 31.8
3. 0.091
4. 513.4
5. 0.2081
6. 2.193
7. 85.003
8. 0.000407

Write each of the following products as a decimal without exponents:

9. $(7.13)(10^2)$
10. $(85.1)(10^{-3})$
11. $(0.0316)(10^3)$
12. $(4,716)(10^{-4})$
13. $(683,000)(10^{-5})$
14. $(0.7508)(10^6)$
15. $(9,073,000)(10^{-7})$
16. $(0.00357)(10^8)$

Perform the indicated operations on each of the following expressions. Simplify your answer completely, writing it without exponents.

17. $(800)(0.005)$
18. $(4.1)(8.03)$
19. $\dfrac{8.4}{0.021}$
20. $\dfrac{0.076}{0.19}$
21. $0.0815 + 9.16$
22. $0.4718 + 0.8509$
23. $19.17 - 12.85$
24. $4.816 - 9.135$
25. $(0.009)(0.36)$
26. $(0.83)(0.00061)$
27. $\dfrac{2{,}400}{0.0006}$
28. $\dfrac{0.056}{7{,}000}$
29. $-7.18 - 3.0914$
30. $-24.008 + 17.8164$
31. $(4.81)10^{-2} + (19.3)10^{-2}$
32. $867 \cdot 10^{-2} + 9{,}142 \cdot 10^{-4}$

5.4. Scientific Notation

In this age of space travel and electronic microscopes, man not only deals with very large numbers, but he is also concerned with very small numbers as well. Consider these practical examples:

(a) The radius of a hydrogen atom is about 0.00000000005 meters, or 0.00000002 inches.
(b) The mass of the earth is 13,200,000,000,000,000,000,000,000 pounds.
(c) About 533,000,000,000,000,000,000,000 molecules are in 1 ounce of oxygen.

In these examples, the numbers are such that they are difficult to write, much less to read. In order to efficiently deal with these numbers, an improved method of writing them had to be devised. The following examples suggest the method which is used:

(d) In (a), $0.00000000005 = \dfrac{5}{10^{11}} = 5 \cdot 10^{-11}$

(e) In (a), $0.00000002 = \dfrac{2}{10^{8}} = 2 \cdot 10^{-8}$

(f) In (b), $13{,}200{,}000{,}000{,}000{,}000{,}000{,}000{,}000 = 1.32 \times 10^{25}$
(g) In (c), $533{,}000{,}000{,}000{,}000{,}000{,}000{,}000 = 5.33 \times 10^{23}$

In examples (f) and (g), the $\times$ is introduced as a symbol for multiplication, instead of the dot, to avoid confusion with the decimal point

Real Numbers 185

when it occurs in the same expression. Since this method of writing very large (or very small) numbers frequently involves decimals, but never algebraic symbols, there should be no confusion over the use of the $\times$ symbol. The student should again be advised that the $\times$ is never used as a symbol for multiplication wherever algebraic symbols will be used.

This method of writing any number as the product of an integral power of ten and a number between 1 and 10, is called **scientific notation.** It is used as a calculating tool in almost every branch of science and engineering. By examining some additional examples of a less imposing nature, this method will soon become obvious:

(h) $0.0007 = \dfrac{7}{10,000}$

$= \dfrac{7}{10^4}$ Observe that the decimal point was moved four places to the right

$= 7 \cdot 10^{-4}$

(i) $0.00081 = \dfrac{81}{100,000}$

$= \dfrac{81}{10 \cdot 10^4}$ Now the decimal point has been moved four places to the right

$= \dfrac{81}{10} \cdot 10^{-4}$

$= 8.1 \times 10^{-4}$

(j) $187 = \dfrac{187 \cdot 10^2}{10^2}$

$= \dfrac{187}{100} \cdot 10^2$ In this example, the decimal point was moved two places to the left

$= 1.87 \times 10^2$

(k) $43,000 = 43 \cdot 10^3$

$= \dfrac{43}{10} \cdot 10^4$ Now the decimal point has been moved four places to the left

$= 4.3 \times 10^4$

186   Pre-Algebra

Notice in (h) and (i), that the resulting number between 1 and 10 is multiplied by a negative power of 10 which is equal to the number of places that the decimal point was moved to the right. In (j) and (k), the resulting number between 1 and 10 is multiplied by a positive power of 10, which is equal to the number of places that the decimal point was moved to the left. These facts can be stated as a general rule.

> In order to write a given number in scientific notation, move the decimal point so that the given number is between 1 and 10. If the decimal point is moved to the right, multiply by 10 raised to a negative power which is equal to the number of places that the decimal point had been moved to the right. If the decimal point is moved to the left, multiply by 10 raised to a positive power which is equal to the number of places that the decimal point had been moved to the left.

Again, some examples which should make this rule clear:

(l) $708 = 7.08 \times 10^2$
(m) $0.00384 = 3.84 \times 10^{-3}$
(n) $48,031 = 4.8031 \times 10^4$
(o) $0.0000867 = 8.67 \times 10^{-5}$

It is also necessary to become familiar with the reverse process, that of changing numbers written in scientific notation back into conventional form. This can easily be performed by noting that multiplication by a positive power of 10 moves the decimal point to the right. Multiplication by a negative power of 10 moves the decimal point to the left. In either case, the power of 10 is the actual number of places that the decimal point is to be moved. The examples below illustrate this idea:

(p) $7.031 \times 10^7 = 70,310,000$
(q) $4.18 \times 10^{-5} = 0.0000418$

Scientific notation is a convenient form with which unwieldy numbers can be written. However, its principal use is in calculations involving these kinds of numbers. Finding the decimal point in the multiplication $0.00006(21,000,000)$ can be difficult. However, by first writing these numbers in scientific notation, and then multiplying, the product can easily be obtained. The following examples illustrate this procedure:

(r) $(0.00006)(21{,}000{,}000) = (6 \cdot 10^{-5})(2.1 \times 10^7)$
$= (6)(2.1)(10^{-5} \cdot 10^7)$
$= 12.6 \times 10^2$
$= 1{,}260$

(s) $(4{,}300{,}000)(0.0000031) = (4.3 \times 10^6)(3.1 \times 10^{-6})$
$= (4.3)(3.1)(10^6 \cdot 10^{-6})$
$= 13.33 \times 10^0$
$= 13.33$

The product of the two decimals, $(4.3)(3.1)$, in example (s), was discussed in the previous section. Now for some examples illustrating calculations which include both multiplication and division:

(t) $\dfrac{(4{,}000)(0.0032)}{0.00064} = \dfrac{(4 \cdot 10^3)(3.2 \times 10^{-3})}{6.4 \times 10^{-4}}$

$= \dfrac{\overset{1}{\cancel{4}}(\overset{2}{\cancel{3.2}}) \cdot 10^3 \cdot 10^{-3}}{\underset{1.6}{\cancel{6.4}} \times 10^{-4}}$

$= 2 \cdot 10^0 \cdot 10^4$
$= 20{,}000$

(u) $\dfrac{(0.028)(0.00003)}{0.0007} = \dfrac{(2.8 \times 10^{-2})(3 \cdot 10^{-5})}{7 \cdot 10^{-4}}$

$= \dfrac{\overset{0.4}{\cancel{2.8}}(3) \cdot 10^{-2} \cdot 10^{-5}}{\underset{1}{\cancel{7}} \cdot 10^{-4}}$

$= (0.4)(3) \cdot 10^{-7} \cdot 10^4$
$= 1.2 \times 10^{-3}$
$= 0.0012$

The numbers in the last two examples have been purposely kept simple. The intent is for the student to learn the method of scientific notation, not to be overly burdened with tedious calculations. When performing calculations which involve difficult multiplications and divisions, various calculating aids are used. These include logarithms, slide rules, adding machines, desk calculators, and computers.

5.4. Exercises

Write each of the following numbers in scientific notation:

1. 0.03
2. 0.0847
3. 800
4. 3,400
5. 0.00731
6. 0.000405
7. 76,000
8. 439,000
9. 0.00004
10. 0.00000081
11. 96,000,000
12. 4,807,000,000
13. 0.05008
14. 0.000000000000503
15. 68,000,000,000,000,000
16. 7,083,500,000,000,000,000

Multiply each of the following indicated products:

17. 4.03×10^{-2}
18. 9.18×10^{-4}
19. 8.05×10^{3}
20. 5.67×10^{5}
21. 6.32×10^{-5}
22. 2.85×10^{-9}
23. 7.8352×10^{12}
24. 3.000085×10^{18}

Perform each of the following calculations by first writing each of the numbers in scientific notation, then simplifying:

25. $\dfrac{(40,000)(0.008)}{800,000}$

26. $\dfrac{(450)(300,000)}{90,000}$

27. $\dfrac{(0.000008)}{(0.00004)(0.005)}$

28. $\dfrac{(81,000,000)(7,000)}{210,000}$

29. $(0.002)(80,000)(4,000,000)(0.000003)$

30. $\dfrac{96,000,000}{(0.008)(0.00003)(0.0002)}$

31. $\dfrac{(31,000)(280)}{35,000,000}$

32. $\dfrac{(2,100)(4,000,000)(0.0003)}{(0.024)(14,000)}$

5.5. Decimal Approximations for Irrational Numbers

The previous sections in this chapter have been primarily concerned with demonstrating how to write numbers in decimal notation. Thus far, only rational numbers have been considered, and it was found that an exact decimal representation can be written for all rational numbers. In this section, an attempt shall be made to find and write decimal representations for irrational numbers. This attempt will not be successful because it will be seen here that it is impossible to write exact decimal representations for irrational numbers, although they exist. It will only be possible to write successive finite approximations to the exact value of a rational number.

The proof that $\sqrt{2}$ cannot be represented by a rational number, and is therefore irrational, lies beyond the scope of this text. The same kind of proof can be used to show that $\sqrt{3}$, $\sqrt{5}$, and $\sqrt{6}$ are irrational. In Section 4.3, p. 142, an irrational number was defined as a number which could not be written as the product of a desired number of equal rational factors. Now it can be proven than an irrational number cannot be represented by the quotient of two integers. It seems reasonable then to define an irrational number as follows:

> An irrational number is one which cannot be written as the quotient of two integers.

In brief, a rational number is one which can be written as the quotient of two integers, and an irrational number is one which cannot.

It was demonstrated in Chapter 4 that the exact value of $\sqrt{2}$ was between $\frac{14}{10}$ and $\frac{15}{10}$. This can be restated with decimals as

$$1.4 < \sqrt{2} < 1.5$$

The following sequence of inequalities is also true:

$1.41 < \sqrt{2} < 1.42$ Since $(1.41)^2 < (\sqrt{2})^2 < (1.42)^2$, and therefore $1.9881 < 2 < 2.0164$

$1.414 < \sqrt{2} < 1.415$ Since $(1.414)^2 < (\sqrt{2})^2 < (1.415)^2$, and therefore $1.999396 < 2 < 2.002225$

$1.4142 < \sqrt{2} < 1.4143$ Since $(1.4142)^2 < (\sqrt{2})^2 < (1.4143)^2$, and therefore
$$1.99996164 < 2 < 2.00024449$$

190 Pre-Algebra

This process can be repeated indefinitely with each step a closer approximation to the exact value. Unfortunately, the process does not yield an exact decimal representation since the only exact value which can be written is $\sqrt{2}$ itself. All other decimal representations are only approximations.

Observe that the decimal approximation of $\sqrt{2}$ does not have a repetition pattern for the four decimal places shown. It can be established that it will not repeat with any pattern, regardless of the number of decimal places which are found. This is true for all irrational numbers. Generally then, an irrational number has a nonterminating, nonrepeating decimal representation.

Another example of a nonrepeating, nonterminating number is π (pi), which is the quotient of the circumference of a circle to its diameter. This number has an interesting history. For centuries, laymen and mathematicians alike had been searching for a rational number with which to express π. Although $\frac{22}{7}$ is sometimes used in elementary mathematical calculations, it is only an approximation, although not a bad one. Expressed to six decimal places, $\frac{22}{7}$ is approximately equal to 3.142857, which is written as $\frac{22}{7} \doteq 3.142857$, and a like approximation for π is 3.141592. In 1961, π was calculated on an IBM 7090 computer to over 100,000 decimal places in slightly less than 9 hours without a repeating pattern occurring. This number is frequently used in all levels of mathematics.

The next question which should arise in the reader's mind is: "How are these decimal approximations determined?" There are several methods by which decimal approximations of irrational numbers can be calculated. The following algorithm is perhaps the most direct and easiest to understand. The algorithm will be demonstrated by finding a decimal approximation for $\sqrt{2}$.

(a) Since $1 < \sqrt{2} < 2$, 1 is a first estimate to $\sqrt{2}$. Now, divide 2 by 1:

$$\begin{array}{r} 2 \\ 1\overline{)2} \\ \underline{2} \\ 0 \end{array}$$

Real Numbers 191

The object is to find two equal factors of 2. The factors 1 and 2 are unequal. A good second estimate is a number halfway between. This is called the **average** of 1 and 2 and it is found by adding the two numbers and then dividing by 2. So the second estimate is $\frac{1+2}{2} = 1.5$. Now divide 2 by 1.5, or since $\frac{2}{1.5} = \frac{20}{15}$, divide 20 by 15:

$$\begin{array}{r} 1.3\ 3 \\ 15\overline{)2\ 0.0\ 0\ 0} \\ \underline{1\ 5} \\ 5\ 0 \\ \underline{4\ 5} \\ 5 \end{array}$$

Again, $1.33 \neq 1.5$, so for a third estimate, use the average of 1.33 and 1.5, which is $\frac{1.33 + 1.5}{2} = 1.415$. Now divide 2 by 1.415, or since $\frac{2}{1.415} = \frac{2,000}{1,415}$, divide 2,000 by 1,415:

$$\begin{array}{r} 1.4\ 1\ 3\ 4\ 2 \\ 1{,}415\overline{)2\ 0\ 0\ 0.0\ 0\ 0\ 0\ 0} \\ \underline{1\ 4\ 1\ 5} \\ 5\ 8\ 5\ 0 \\ \underline{5\ 6\ 6\ 0} \\ 1\ 9\ 0\ 0 \\ \underline{1\ 4\ 1\ 5} \\ 4\ 8\ 5\ 0 \\ \underline{4\ 2\ 4\ 5} \\ 6\ 0\ 5\ 0 \\ \underline{5\ 6\ 6\ 0} \\ 3\ 9\ 0\ 0 \\ 2\ 8\ 3\ 0 \end{array}$$

Again, $1.415 \neq 1.41342$, so for a fourth and final estimate, average 1.415 and 1.41342, which is $\frac{1.415 + 1.41342}{2} = 1.41421$. Now divide 2 by 1.41421, or since $\frac{2}{1.41421} = \frac{200{,}000}{141{,}421}$, divide 200,000

192   Pre-Algebra

by 141,421:

$$\require{enclose}\begin{array}{r}1.4\,1\,4\,2\,1\,7\\141{,}421\enclose{longdiv}{2\,0\,0\,0\,0\,0.0\,0\,0\,0\,0\,0}\\1\,4\,1\,4\,2\,1\\\hline 5\,8\,5\,7\,9\,0\\5\,6\,5\,6\,8\,4\\\hline 2\,0\,1\,0\,6\,0\\1\,4\,1\,4\,2\,1\\\hline 5\,9\,6\,3\,9\,0\\5\,6\,5\,6\,8\,4\\\hline 3\,0\,7\,0\,6\,0\\2\,8\,2\,8\,4\,2\\\hline 2\,4\,2\,1\,8\,0\\1\,4\,1\,4\,2\,1\\\hline 1\,0\,0\,7\,5\,9\,0\\9\,8\,9\,9\,4\,7\\\end{array}$$

Observe that $1.41421 \neq 1.414217$, but these approximations are correct to five decimal places. It can easily be verified that

$$(1.41421)^2 = 1.9999899241$$

If greater accuracy is desired, it is only necessary to continue the algorithm to the desired accuracy. Therefore, $\sqrt{2} \doteq 1.41421$.

A second example may help in understanding the process.

(b) To find $\sqrt{10}$, observe that $3 < \sqrt{10} < 4$, so 3 is used as a first estimate. Therefore, divide 10 by 3:

$$\begin{array}{r}3.3\,3\\3\enclose{longdiv}{1\,0.0\,0\,0}\\9\\\hline 1\,0\\9\\\hline 1\,0\\\end{array}$$

The average of 3.33 and 3 is $\dfrac{3 + 3.33}{2} = 3.165$. Since $\dfrac{10}{3.165} = \dfrac{10{,}000}{3{,}165}$, divide 10,000 by 3,165:

$$\begin{array}{r} 3.1595 \\ 3{,}165 \overline{) 10000.0000} \\ 9495 \\ \hline 5050 \\ 3165 \\ \hline 18850 \\ 15825 \\ \hline 30250 \\ 28495 \\ \hline 17550 \\ 15825 \\ \hline \end{array}$$

The third estimate is the average of 3.165 and 3.1595, which is $\frac{3.165 + 3.1595}{2} = 3.1622$. Since $\frac{10}{3.1622} = \frac{100{,}000}{31{,}622}$, divide 100,000 by 31,622:

$$\begin{array}{r} 3.16235 \\ 31{,}622 \overline{) 100000.00000} \\ 94866 \\ \hline 51340 \\ 31622 \\ \hline 197180 \\ 189732 \\ \hline 74480 \\ 63244 \\ \hline 112360 \\ 94866 \\ \hline 174940 \\ 158110 \\ \hline \end{array}$$

The average of 3.1622 and 3.16235 is $\frac{3.1622 + 3.16235}{2} = 3.162275.$

Now for a fourth and final estimate, divide 10,000,000 by 3,162,275:

$$\begin{array}{r}3.1\,6\,2\,2\,8\,0\\3{,}162{,}275\overline{)1\,0\,0\,0\,0\,0\,0\,0.0\,0\,0\,0\,0\,0}\\9\,4\,8\,6\,8\,2\,5\\\hline 5\,1\,3\,1\,7\,5\,0\\3\,1\,6\,2\,2\,7\,5\\\hline 1\,9\,6\,9\,4\,7\,5\,0\\1\,8\,9\,7\,3\,6\,5\,0\\\hline 7\,2\,1\,1\,0\,0\,0\\6\,3\,2\,4\,5\,5\,0\\\hline 8\,8\,6\,4\,5\,0\,0\\6\,3\,2\,4\,5\,5\,0\\\hline 2\,5\,3\,9\,9\,5\,0\,0\\2\,5\,2\,9\,8\,2\,0\,0\\\hline 1\,0\,1\,3\,0\,0\,0\end{array}$$

Here, $\sqrt{10} \doteq 3.162275$, which is accurate to 5 decimal places. Obviously, this method is lengthy, yet it is as easy as any if the square root must be found without any additional aid. Fortunately, there are a number of more rapid ways to find these values, the most common of which is a table. A table of squares and square root approximations, accurate to three decimal places, appears on p. 267. Tables are also available having square root approximations in excess of three decimal places. When evaluating lengthy and complicated problems, logarithms, slide rules, desk calculators, or computers are used, depending upon the accuracy and speed desired.

Some irrational numbers can be easily evaluated by first simplifying the radical. The following examples will illustrate this idea:

(c) $\sqrt{8} = 2\sqrt{2}$
       $\doteq 2(1.4142)$ From example (a)
       $\doteq 2.8284$

(d) $\sqrt{800} = \sqrt{100}\sqrt{8}$
           $= 10 \cdot 2\sqrt{2}$
           $= 20\sqrt{2}$
           $\doteq 20(1.4142)$
           $\doteq 28.284$

(e) $\sqrt{1{,}000} = \sqrt{100}\sqrt{10}$
$= 10\sqrt{10}$
$\doteq 10(3.1624)$ From example (b)
$\doteq 31.624$

(f) $\sqrt{40} = 2\sqrt{10}$
$\doteq 2(3.1624)$
$\doteq 6.3248$

Actually, square root tables need not go beyond values up to 100 because of the principle which these examples illustrate.

It was established in this section that it is impossible to write the exact decimal representation of an irrational number. Instead, decimal approximations can be written, to any finite number of places desired. The decimal representation of an irrational number will be nonrepeating, and nonterminating. Also included in this section was an algorithm with which the decimal approximation can be found for the square root of non-perfect square rational numbers to any finite number of places.

5.5. Exercises

Using the square root algorithm, find the decimal approximations to three decimal places for each of the following irrational numbers

1. $\sqrt{6}$
2. $\sqrt{7}$
3. $\sqrt{11}$
4. $\sqrt{13}$
5. $\sqrt{14}$
6. $\sqrt{29}$
7. $\sqrt{35}$
8. $\sqrt{4.9}$

If $\sqrt{2} \doteq 1.41421$, $\sqrt{3} \doteq 1.73205$, and $\sqrt{5} \doteq 2.23607$, then evaluate each of the following, accurate to four decimal places.
(Example: $\sqrt{2} = \sqrt{4}\sqrt{3} = 2\sqrt{3} \doteq 2(1.7321) \doteq 3.4642$)

9. $\sqrt{20}$
10. $\sqrt{18}$
11. $\sqrt{32}$
12. $\sqrt{200}$
13. $\sqrt{98}$
14. $\sqrt{45}$
15. $\sqrt{27}$
16. $\sqrt{180}$

5.6. Real Number System

It was demonstrated in the previous section that an irrational number has a nonterminating, nonrepeating decimal representation. Referring again to Chapter 4, it can be seen that the decimal representation of rational numbers are either terminating, or nonterminating with a repeating pattern. The union of these two sets of numbers produces a new set which ideally possesses at least the properties of its subset, the rational numbers. It is called the **real number system** and it is usually designated by the letter R.

Since all elements of Q, the set of rational numbers, are also elements of R, $Q \subseteq R$. The elements of H, the set of irrational numbers, are also elements of R, so that $H \subseteq R$. Yet, Q and H are disjoint sets. The set of integers, J, can be written $\{\ldots, -3.\bar{0}, -2.\bar{0}, -1.\bar{0}, 0.\bar{0}, 1.\bar{0}, \ldots\}$, so it should be obvious that this set is also a subset of R, but particularly of Q, since these are repeating, nonterminating decimals. Stated symbolically, $J \subseteq R$ and $J \subseteq Q$.

Since the whole numbers are also integers, then $W \subseteq J$, and all whole numbers must be elements of R, so that $W \subseteq R$. Finally, since $N \subseteq W$, and the natural numbers are a subset of the whole numbers, then it follows that $N \subseteq R$. The structure of the real number system can be illustrated with the chart shown at the top of p. 197.

The real number system includes all the numbers which are used in arithmetic. A number which can be either rational or irrational, but not both, is called a **real number.** In more advanced mathematics, there are other numbers which will supplement this system, but the real number system is quite sufficient to carry the student through elementary algebra.

Real numbers which are positive include the positive integers, positive rationals, and positive irrational numbers, all of which are subsets of the set of real numbers. Numbers in each of these sets are to the right of zero, as illustrated in the diagram below:

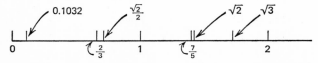

The inequality $n > 0$, n a real number, includes all real numbers which are to the right of zero on the number line—these are called the positive real numbers. The inequality $n < 0$, n a real number, includes all real numbers which are to the left of zero on the number line—these are called the negative real numbers.

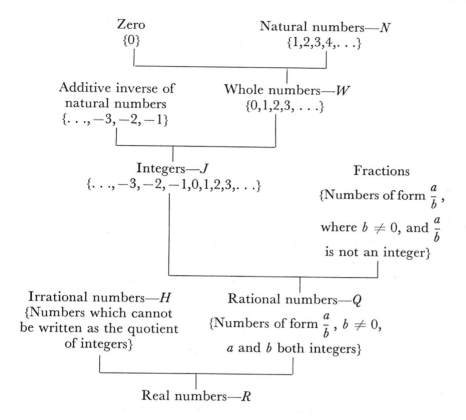

It would be useful at this point to review the properties of the real number system. The real number system:

1. Is closed with respect to addition and multiplication.
2. Is commutative with respect to addition and multiplication.
   $$a + b = b + a$$
   $$a \cdot b = b \cdot a$$
   where a and b are any two real numbers
3. Is associative with respect to addition and multiplication.
   $$(a + b) + c = a + (b + c)$$
   $$(a \cdot b) \cdot c = a \cdot (b \cdot c)$$
   where a, b, and c are any three real numbers
4. Has identity elements with respect to addition and multiplication.
   0 is the identity element for addition
   1 is the identity element for multiplication

5. Has inverse elements with respect to addition and multiplication.
   $-a$ is the additive inverse of a
   $\dfrac{1}{a} = a^{-1}$ is the multiplicative inverse of a, $a \neq 0$
   In both cases, a is a real number.
6. Has a distributive property.
   $a \cdot (b + c) = a \cdot b + a \cdot c$ for a, b, and c any three real numbers
   $(b + c) \cdot a = b \cdot a + c \cdot a$

These six statements, which include eleven properties, are called the **field properties** of a number system. All number systems satisfying these properties are called **fields**. In particular, the real number system is a field.

The real number system has the following properties as well:

7. Any two elements, or numbers, in the real number system can be ordered. That is, given any two real numbers, a and b, $a > b$, $a = b$, or $a < b$, where only one of these three statements can be true at one time. Thus, the real number system is an **ordered field**.
8. The real number system is **dense**. That is, given any two real numbers, there is at least one, and actually infinitely many real numbers between them on the number line.

It is almost a certainty that some readers would have preferred to see more of the properties proven instead of assumed. This is a healthy and encouraging sign of a desire to learn more mathematics. The degree of rigor, as this level of preciseness is called, will continue to improve with each mathematics course attempted. The author would be content if the student can retain and properly use the properties which have been developed in this text.

5.6. Exercises

Supply the name of the "smallest" set which makes each of the following statements correct. Use the set designations N, W, J, Q, H, and R:

1. $\{3, 8, 17\} \subseteq \ldots$
2. $\left\{\dfrac{1}{2}, \dfrac{-7}{4}, \dfrac{17}{13}\right\} \subseteq \ldots$

3. $\left\{4, \dfrac{-3}{8}, 7\right\} \subseteq \ldots$

4. $\{-8, 4, 13\} \subseteq \ldots$

5. $\left\{\dfrac{-5}{7}, -3, 0.84\right\} \subseteq \ldots$

6. $\{4.38, -7.2, 9.38\} \subseteq \ldots$
7. $\{4, 9, 13\} \subseteq \ldots$
8. $\{-\sqrt{25}, -8, \sqrt{3}\} \subseteq \ldots$
9. $\{7, 18, 0\} \subseteq \ldots$

10. $\left\{\dfrac{8}{2}, \dfrac{-9}{3}, -3\right\} \subseteq \ldots$

11. $\left\{\dfrac{\sqrt{16}}{2}, \dfrac{\sqrt{36}}{3}, \dfrac{\sqrt{64}}{4}\right\} \subseteq \ldots$

12. $\{\sqrt{17}, \sqrt{23}, \sqrt{27}\} \subseteq \ldots$
13. $\{7, 8, 14\} \cup \{-3, -12, -17\} \subseteq \ldots$

14. $\left\{-3, \dfrac{1}{2}, \dfrac{-8}{3}\right\} \cup \{\sqrt{3}\} \subseteq \ldots$

15. $\left\{\dfrac{-7}{4}, \dfrac{-13}{7}, \dfrac{8}{5}\right\} \cup \{0.783, -4.3\overline{7}, 17\} \subseteq \ldots$

16. $\{-7, -4, 23\} \cup \{14, 27, \pi\} \subseteq \ldots$

5.7. Summary

The principal function of this chapter was to develop decimal notation. In order to do this, it was first necessary to introduce the negative exponent. With it, the concept of writing numbers in positional notation can be extended to numbers whose values are less than 1. The introduction of the zero exponent was desirable to be able to show the natural continuity of exponents from the natural numbers, through zero, to the negatives of the natural numbers. In other words, to demonstrate that exponents can be extended from natural numbers to integers, by way of simple definitions.

Pre-Algebra

With the new decimal notation, it was then possible to demonstrate that all numbers have a decimal representation. Specifically, the decimal representation of a rational number may be terminating, or repeating and nonterminating, depending upon whether the denominator is a power of 10, a factor of a power of 10, or neither. A number which is irrational, has a nonrepeating, nonterminating decimal representation.

Finally, this new set of numbers with its properties, was called the real number system. Elements of this set are referred to as real numbers. The properties of the real number system and the relationship between its subsets were discussed.

5.7. Review Exercises

Express each of the following with a single positive exponent:

1. 7^{-4}
2. $\dfrac{1}{5^{-3}}$
3. -4^{-3}
4. $\left(\dfrac{-4}{3}\right)^{-4}$
5. $\dfrac{7^0}{-8^{-3}}$
6. $5^{-3} \cdot 5^{-4}$
7. $9^2 \cdot 9^{-5}$
8. $\dfrac{6^2}{6^{-4}}$

Simplify each of the following expressions so that each numeral is represented only once in exponential notation. Express your answer so that it is not in fractional form.

9. $\dfrac{3^4}{3^7}$
10. $\dfrac{(5^2 + 3)^0}{5^{-3}}$
11. $\dfrac{4^{-3}(7 + 4)^0}{4^5}$
12. $\dfrac{5^3}{5^{-6} \cdot 5^5}$
13. $\dfrac{4^0}{7^2 7^{-8}}$
14. $(4^3 \cdot 5^{-8} \cdot 8^0)^4$
15. $\dfrac{6^3}{4^7 \cdot 6^{-3}}$
16. $\left(\dfrac{5^{-2}}{4^5}\right)^{-6}$

Write each of the following in decimal notation:

17. $\dfrac{377}{10{,}000}$
18. $\dfrac{79}{100}$
19. $\dfrac{809}{1{,}000}$
20. $\dfrac{23}{100{,}000}$

Real Numbers 201

Write each of the following decimals as a common fraction whose denominator is a power of 10:

21. 0.0837   22. 0.6397   23. 0.00037   24. 0.43573

Convert each of the following fractions into decimal fractions:

25. $\dfrac{7}{20}$ 27. $\dfrac{7}{16}$ 29. $\dfrac{5}{11}$ 31. $\dfrac{49}{400}$

26. $\dfrac{5}{12}$ 28. $\dfrac{13}{80}$ 30. $\dfrac{11}{18}$ 32. $\dfrac{5}{27}$

Write each of the following decimals as the product of a natural number and a power of 10:

33. 0.07   34. 26.53   35. 0.3405   36. 0.00008215

Write each of the following products as a decimal without exponents:

37. 0.00734×10^4 39. 49.78×10^4
38. $9{,}216 \cdot 10^{-5}$ 40. 0.2739×10^{-3}

Perform the indicated operations on each of the following expressions. Simplify your answers completely, writing it without exponents:

41. $(0.007)(6{,}000)$ 45. $(0.48)(0.0012)$

42. $\dfrac{96}{0.008}$ 46. $-21.38 + 7.84$

43. $4.83 - 2.37$ 47. $9.73 \times 10^{-3} + 4.48 \times 10^{-2}$

44. $\dfrac{0.0048}{0.12}$ 48. $-8.04 \times 10^{-2} - 6.98 \times 10^{-3}$

Write each of the following numbers in scientific notation:

49. 0.0081   53. 0.00006007
50. 6,800   54. 84,200,000
51. 0.78   55. 0.0007035
52. 42,750   56. 2,386,000,000,000

Write each of the following products without exponents:

57. 7.4×10^{-3} 58. 9.821×10^5 59. 3.003×10^{-6} 60. 5.267×10^7

Perform each of the following calculations by first writing each of the numbers in scientific notation, then simplifying:

61. $\dfrac{24{,}000(0.02)}{0.0004}$

62. $(0.05)(48{,}000)(0.00006)(5{,}000{,}000)$

63. $\dfrac{(42{,}000{,}000)(50{,}000)}{700}$

64. $\dfrac{0.00006}{(0.0008)(0.005)}$

Find the decimal approximation correct to three decimal places for each of the following irrational numbers:

65. $\sqrt{15}$ 66. $\sqrt{59}$ 67. $\sqrt{8.1}$ 68. $\sqrt{17.4}$

If $\sqrt{2} \doteq 1.41421$, $\sqrt{3} \doteq 1.73205$, and $\sqrt{5} \doteq 2.23607$, then evaluate each of the following, accurate to four decimal places:

69. $\sqrt{125}$ 70. $\sqrt{128}$ 71. $\sqrt{108}$ 72. $\sqrt{147}$

6 Mathematical Sentences

6.1. Algebraic Expressions

Now that the real number system has been developed, it would be interesting and useful for the student to see how these numbers are used in algebra. First, what is algebra? Every year, this instructor takes delight in asking advanced mathematics students this question, because it is a rare student who can supply a meaningful reply to this seemingly simple question. Actually, there are many different kinds of algebra. For our purposes though, algebra is merely a generalization of arithmetic. Using algebra, problems can be solved which normally cannot be with arithmetic.

It is not the purpose of this text to study algebra, but the student planning to do so will find it helpful to first become acquainted with its language. When solving a problem with the use of algebra, the numerical facts are first written as **mathematical sentences** or **mathematical statements.** These mathematical sentences can then be manipulated by algebraic methods. If sufficient information is given, the desired answer to the problem can be found. This chapter shall attempt to provide the reader with some experience in converting written or verbal statements into equivalent statements which include mathematical symbols only. It shall also acquaint the reader with some of the mathematical terms that are associated with this topic. The algebraic methods shall come in algebra, which normally follows this course.

Before discussing mathematical sentences, it is desirable to first become familiar with **mathematical expressions.** A mathematical expression can be compared to a grammatical phrase in a sentence. That is, it is a group of words which functions as a unit of meaning in a sentence. Examples of grammatical phrases are "the green house," "the quiet girl," and "up the steep hill." The following are examples of mathematical phrases or mathematical expressions. Note that they form a unit of meaning but are not complete sentences or problems.

(a) The cube of five
(b) Twice the sum of four and five
(c) The sum of twice four and three

Examples (a), (b), and (c) can be replaced by specific numerals and other symbols to become (a) 5^3, (b) $2(4 + 5)$, and (c) $2 \cdot 4 + 3$. These mathematical expressions have been translated into mathematical symbols in the same way that any statement is translated from one language to another. Please observe that no request was made to find the cube of five in (a), or to find the sums in (b) and (c). That is, they are merely mathematical phrases and not complete mathematical sentences or problems. It is important that the student learns to translate a given statement into symbols with as little change as possible. Mathematics is a language whose symbolism must be understood and used correctly no less than the written symbolism (words) of any other language.

The shorthand which a secretary utilizes is also a language which resembles mathematics insofar as strange-looking symbols are used to represent a word or a collection of words. In illustration, 2 is the mathematical symbol for two, while $\mathcal{P}$ is the shorthand symbol for two. Who can judge which is the stranger of the two symbols?

Frequently, the mathematician wishes to work with a quantity which is not known. He does this by letting the unknown quantity be represented by a letter. That is, he may speak of d dollars, p pounds, or x miles. The choice of the letter which is used is relatively unimportant. It is your author's conjecture that x is popular because it is an easy symbol to make, and it is not confused with any other letter of the alphabet. Consider these examples:

(d) Bob and Linda have a certain number of dollars.
(e) Mel and Zel traveled so many miles last week.
(f) The distance between Cindy and Sharon.

Any letter can be used to represent the unknown quantities here. For example, let r be the number of dollars in (d), let s be the number of miles in (e), and let t be the number of feet between the girls in (f). Each of these letters represent the quantities involved. Since they each represent only one quantity, they are called **constants.** Other examples of constants are 7, $\frac{23}{47}$, $\sqrt{3}$, and π.

Letters can also represent unknowns which are not numerical quantities. The following examples illustrate these kinds of situations:

(g) He is a student at American River College.
(h) It is a four-legged animal.
(i) It is a green book in the Library of Congress.

In (g), let x be the representation for a male student at American River College. Let y represent a four-legged animal in (h), and let z be a green book in (i). Here, each of x, y, and z are representations for many "values" or replacements. Since x, y, or z can have various replacements, they are called **variables,** and the set of replacements for each of these variables is called a **replacement set.** For example, the replacement set for x in (g) is all males who are students at American River College. In (h), $y = \{$horse,cow,lion,beaver,...$\}$. Here, the replacement set is finite, but the elements are too numerous to list. In (i), z represents a finite set, since it is a subset of the finite set of books in the Library of Congress.

The replacement set for a variable is usually called the **domain** of the variable. When the domain of a variable is a one-element set, the variable is called a constant. When the domain of a variable is numerical in nature, each of the elements of the replacement set is usually referred to as the value of the variable. A variable can take on many values. Again then, a variable having just one value is called a constant.

Now consider the following examples, which include both constants and variables, and their equivalent mathematical expressions:

(j)   The sum of a given number and two $x + 2$
(k)   Twice a given number $2 \cdot y$
(l)   Four times a given number less seven $4 \cdot z - 7$

The multiplication dots in (k) and (l) are usually omitted and then simply written as $2y$ and $4z - 7$. That is, when a constant and a variable, or two variables are written next to each other without a sign between them, it is understood that they are to be multiplied. It should be clear that when two or more constants are written next to each other, they do not represent multiplication. Specifically, 43 represents forty-three and not four times three. Again, the choice of letters in the above was arbitrary.

The domains of the variables in the examples above are not known since insufficient information was supplied. Usually, though not always, the nature of the problem will determine the domain of the variable. The following problems will illustrate this statement:

(m) Bob has twice as many dollar bills as Karen, and together they have $87.00. Find the number of dollar bills that Bob has.
(n) The sum of two consecutive integers is -43. Find the integers.
(o) The lengths of the two legs of a right triangle are 3 and 5 inches. Find the length of the hypotenuse.

The number of dollar bills that Bob has must certainly be restricted to a natural number. Therefore, the domain of the variable in (m) is N, the set of natural numbers. On first glance, the domain of (n) is J, the set of integers. However, upon reflection, one can determine that the domain can be restricted to the negative integers. **Consecutive** integers are those integers which follow each other on the number line. Examples of consecutive integers are 7, 8 and $-4, -3$. In (o), the hypotenuse of the triangle must certainly be a positive number, although it is not restricted to a natural number. If x denotes the length of the hypotenuse, the domain of this variable will be in the set of positive real numbers.

Mathematical Sentences

Mathematical expressions which contain numerical quantities only are referred to as **numerical expressions.** Any mathematical expression which contains numerical quantities only, variable quantities only, or variable and numerical quantities is called an **algebraic expression.** Examples of numerical expressions are $3 \cdot 2 + 7$, $\dfrac{5 \div 3 - 8}{4}$, and $2(4 \cdot 3 - 5)$. Examples of algebraic expressions are $7 \cdot 4 - 8 \cdot 3$, $2x - 1$, $3(4x + 7)$, and $\dfrac{3x + 2y}{5z}$.

Most symbols of mathematics can be represented by at least one and usually several different expressions. The expressions, "and," "more than," or "increased by" are usually equivalent to plus or addition. The subtraction operation is usually represented by one of the expressions: "less," "less than," or "decreased by." Although "less" and "less than" sound almost the same, they do not have the same meaning. Seven less two is written as $7 - 2$, while seven less than two is symbolized as $2 - 7$.

6.1. Exercises

Translate, exactly as stated, each of the following statements into numerical expressions. Do not perform the indicated operations:

1. The square of six
2. The negative of four
3. The difference of nine and two
4. The product of eight and three
5. The quotient of seven and five
6. The sum of negative four and seven
7. Two less eight
8. Two less than eight
9. The sum of twice seven and eight
10. Twice the sum of seven and eight
11. Twice nine less three times eight
12. Twice the difference of nine and three times eight
13. The negative of the sum of four and seven
14. The quotient of seven and three plus two

15. The quotient of seven and three, plus two
16. The negative of eight square
17. The square of negative eight
18. The cube of the sum of three and five
19. The sum of the cubes of three and five
20. The square of the sum of the cubes of two and five
21. The sum of the squares of two and five cubed
22. The square root of the sum of four and seven
23. The sum of the square roots of four and seven
24. The cube root of the square of the difference of eight and three

Translate, exactly as stated, each of the following into algebraic expressions. Use the letters $x, y,$ and z, in the order stated, whenever one, two, or three variables are required:

25. Twice a given number
26. Six more than a given number
27. The product of two given numbers (not necessarily equal)
28. The square of a number less the cube of a second number
29. A given number less eight
30. Eight less a given number
31. Three less than twice a given number
32. Three less twice a certain number
33. Three times the square of a certain number less nine
34. Three times the square of a given number subtracted from nine
35. Negative three greater than four times a given number
36. Seven less than negative four times the square of an unknown quantity
37. The square of the difference of two numbers
38. The difference of the squares of two numbers
39. Three times a certain number less four times a second
40. The sum of twice the cube of a given number and three times the square of a second
41. Three times the square of the sum of two numbers
42. The sum of twice a given number, three times a second, and five times a third number
43. The product of a given number with the square of a second number

Mathematical Sentences

44. The square of the product of two given numbers
45. Three times the product of the square of one number and the cube of a second
46. The cube of the product of three times a given number with a second number
47. The sum of twice the cube of a certain number and six times that same number, less eleven
48. The sum of twice the cube of a certain number and six times that same number less eleven

6.2. Applications of Algebraic Expressions

Much of the mathematics requires the ability to accurately translate statements into algebraic expressions. It would be well to have some additional practice doing this before continuing. The reader should study the practical exercises in this section thoroughly, since they form the foundation for the problem-solving techniques which are a vital part of algebra.

Algebra was defined in the previous section as the generalization of arithmetic. A few examples which arrive at generalizations should help clarify this statement:

(a) If pencils cost 5 cents each,

2 pencils	cost	$2 \cdot 5$	cents
7 pencils	cost	$7 \cdot 5$	cents
327 pencils	cost	$327 \cdot 5$	cents

Any quantity of pencils, or r pencils, cost

$$r \cdot 5 \quad \text{cents}$$

which is better stated as

$$5r \quad \text{cents}$$

(b) If traveling at 45 miles per hour,

In	2 hours,	you have traveled	$2 \cdot 45$	miles
In	9 hours,	you have traveled	$9 \cdot 45$	miles
In	37 hours,	you have traveled	$37 \cdot 45$	miles

In any number of hours, or t hours, you have traveled

$$t \cdot 45 \quad \text{miles}$$

which is better stated as

$$45t \quad \text{miles}$$

(c) Three consecutive integers are 6, 7, and 8, which can be stated as 6, $6+1$, and $6+2$. Again, three consecutive integers are -5, -4, and -3, which can be stated as -5, $-5+1$, $-5+2$. Generally then, any three consecutive integers can be stated as

$$x, x+1, x+2$$

(d) The number which has a tens digit of 2 and a units digit of 3 can be written as $2 \cdot 10 + 3$. The number which has a tens digit of 4 and a units digit of x can be written as $4 \cdot 10 + x$. The number which has a tens digit of y and a units digit of 7 can be written as $10 \cdot y + 7$, or $10y + 7$.

Observe that the numerical quantities in each example could easily be generalized after a pattern was established. If the reader becomes confused when dealing with a difficult generalization, he will find it helpful to first determine a pattern to assist him in much the same way as was done in the above examples. To illustrate this point, find an expression for the cost of r pencils costing s cents each. Referring to example (a), the reader should observe that the cost of r pencils at 5 cents each resulted in the product of the number of pencils by the cost per pencil. Using the same principle, the cost of r pencils costing s cents each is the product $r \cdot s$, or simply rs.

The ability to generalize numerical quantities comes with practice. Students having difficulty with this kind of work will be rewarded by working as many exercises as possible.

6.2. Exercises

Find a suitable algebraic expression which most nearly expresses each of the following statements:

1. If a represents the age of Terrie, express the age of Emma, who is three times as old.

2. If h represents the height of Gary, express the height of Taylor, who is half as tall as Gary.
3. If a rectangular piece of ground is x feet long and its width is three times as long as its length, express its width in terms of x.
4. If Sara is y years old:
   (a) Represent her age 5 years from now.
   (b) Represent her age 4 years ago.
5. If Larry is x years old:
   (a) Express Pete's age if he is 2 years older.
   (b) Express Debbie's age if she is 5 years younger.
6. If 46 pencils are laid side by side, and each pencil is p inches wide, express the total width in terms of p.
7. Mr. Stewart bought a box of plums which contained eight rows. How many plums did he buy if each row contained p plums?
8. If x is an integer, represent the next larger integer in terms of x.
9. If x is an integer, represent the next smaller integer in terms of x.
10. If x is an even integer, represent the next smaller even integer in terms of x.
11. If x is an even integer, represent the next larger even integer in terms of x.
12. If x is an odd integer, represent the next larger odd integer in terms of x.
13. If x is an odd integer, represent the next smaller odd integer in terms of x.
14. If x is an odd integer, express the next two consecutive odd integers in terms of x.
15. If x is an even integer, express the next two consecutive even integers in terms of x.
16. Represent all the multiples of 3 in terms of y, a natural number.
17. Find an expression for the distance traveled after going t hours at a rate of r miles per hour. [See example (b).]
18. Find an expression for any two-digit number which has a tens digit of t and a units digit of u. [See example (d).]
19. Frances has x dollars and Alice has 6 dollars more than Frances.
    (a) Find an expression for the amount of money that Alice has.

(b) Find an expression for the total amount of money that the two have.
20. Find an expression for the number of pennies in x dollars.
21. Find an expression for the number of nickels in x dollars.
22. Find an expression for the number of dimes in x dollars.
23. Find an expression for the number of quarters in x dollars.
24. Find an expression for the number of half-dollars in x dollars.
25. Find an expression for the number of pennies in n nickels.
26. Find an expression for the number of pennies in d dimes.
27. Find an expression for the number of pennies in q quarters.
28. Find an expression for the number of nickels in d dimes.
29. Find an expression for the number of dimes in n nickels.
30. Find an expression for the number of dimes in h half-dollars.
31. Find an expression for the number of dimes in q quarters.
32. Find an expression for the number of inches in r feet.
33. Find an expression for the number of hours in m minutes.
34. Find an expression for the number of feet in y yards.
35. Find an expression for the number of dollars in n nickels.
36. Find an expression for the number of dollars equivalent to d dimes.
37. Find an expression for the number of dollars equivalent to q quarters.
38. Find an expression for the number of dollars equivalent to $(x+3)$ dimes and $(y-7)$ quarters.
39. A badminton court is 6 feet longer than twice its width.
    (a) If w represents the width, express the length of the court in terms of w.
    (b) If l represents the length, express the width of the court in terms of l.
40. Mary is 20 years less than twice the age of Pearl. If Mary's age is represented by m, express Pearl's age in terms of m.
41. Max is 26 years younger than twice the age of Duke.
    (a) Represent Max's age if Duke is x years old.
    (b) Represent each of their ages 4 years ago.
    (c) Represent Duke's age if Max is y years old.
42. Faris' age is twice Leona's. Find an expression for the sum of their ages if x represents Leona's age.

43. Let *s* represent the length of one side of a square.
    (a) Find an expression for the perimeter of the square.
    (b) Find an expression for the area of the square.
44. Let *l* represent the length and *w* the width of a rectangle.
    (a) Find an expression for the perimeter of the rectangle.
    (b) Find an expression for the area of the rectangle.
45. Given a rectangle whose length is 3 inches less than twice its width. Letting *l* represent the length:
    (a) Find an expression for its perimeter.
    (b) Find an expression for its area.
46. Given a rectangle whose length is $(x + 3)$ inches and whose width is $(y - 7)$ inches,
    (a) Write an expression for the perimeter of the rectangle.
    (b) Find an expression for the area of the rectangle.
47. Given a triangle where one side is 4 inches less than a second side. The third side is twice the shortest side. Express the lengths of each of the sides of the triangle in terms of *x*, the length of the shortest side.
48. It takes *x* hours to mow a lawn.
    (a) Write an expression for the length of time that it takes to mow $\frac{1}{4}$ of the lawn.
    (b) Write an expression for the fraction of the lawn which is mowed in 1 hour.
    (c) Write an expression for the fraction of the lawn which is mowed in 3 hours.

6.3. Equations

Algebraic expressions, which were discussed in Section 6.2, are very much like phrases in grammar. That is, they do not convey a complete thought. Nothing can be said about the value of the mathematical phrase, $x + 3$, until the value of *x* or $x + 3$ is known. Likewise, nothing more can be said about the descriptive phrase, "the old green house." The old green house may be haunted, burning, or sold. Only when two or more grammatical phrases which include a verb are connected, will a sentence be formed containing a complete thought. In much the same way, only when two or more mathematical phrases, or algebraic expressions, are connected with the mathematical equivalent of a verb, will a mathematical sentence be formed. The mathematical relation, discussed in Section 1.7, serves as the verb in a mathematical sentence. That is, the mathematical relation connects two mathematical phrases to form a complete thought. In particular, an equality can be defined as two algebraic expressions which are related by an equals sign. The following are examples of mathematical equalities:

(a) $6 + 3 = 12$ (b) $4 + 5 = 9$ (c) $x + 6 = 8$

Each of the three examples above are sentences, since they contain two algebraic expressions which are separated by an equals sign. Example (a) is a false sentence, while example (b) is true. Example (c) is neither true nor false, but depends upon the proper replacement of x to make the statement true or false. Observe in (c), that when x is replaced by 4, the statement is false, and when x is replaced by 2, the sentence is true. This statement which is neither true nor false, but depends upon the choice of the replacements of the variable, is called an **open sentence**. Now consider these examples to further illustrate this idea:

(d) $9 = 7 - x$ (e) $2y - 1 = 0$ (f) $4z - 1 = 2z + 3$

Observe that these sentences consist of algebraic expressions on both sides of the equals sign. The student should recognize that the 0 on the right side of the equality in (e) is a numerical expression and therefore an algebraic expression.

In (d), the replacement of x by 3 and any other number except -2 results in a false sentence. However, the replacement of x by -2 makes the sentence true. In (e), replacing y by 4 makes the statement false, while replacing y by $\frac{1}{2}$ makes the sentence true. Again in (f), replacing z by 3 makes the sentence false, while a true sentence results when z is replaced by 2. These three examples are all called open sentences.

The domain of the variable sometimes determines whether the statement can be true at any time. Consider example (e) to illustrate this point. Example (e) is true only when the value of y is $\frac{1}{2}$. If the domain of y is the set of natural numbers, then this sentence can never be true, since $\frac{1}{2}$ is not a natural number, and therefore, y can never have that value.

The value or values of the variable which result in a true sentence are called the **solutions** or **roots** of an open sentence. It is also said that a solution or root **satisfies** a sentence. Equalities which are open sentences are generally called **equations.** These equations depend upon the proper replacements of the variable, or solutions to make the sentence true. For this reason, they are called **conditional** equations, but they are generally referred to as simply "equations."

Some conditional equations have only one solution, while others may have many, or a set of solutions. For this reason, the complete solution of an equation is called its **solution set,** even though that set may only have one element, or even no elements. Some sentences are never true and therefore have the null set as their solution set. Solution sets are sometimes referred to as **truth sets** because these are the replacement values of the variable which makes the given sentence true. When an equation is always true, for all replacements of the variable in a given domain, that equation is called an **identity.**

The following equations and their domains should serve to make the previous terms clear:

Equation		Domain
(g) $2x + 6 = 2(x + 3)$	R,	the set of real numbers
(h) $4x - 1 = 7$	N,	the set of natural numbers
(i) $2x + 1 = -5$	W,	the set of whole numbers

It can be verified that all members of R satisfy (g). That is, all members of R make the sentence true and therefore are solutions of (g). For this reason, (g) is an identity. In (h), $x = 2$ satisfies the equation, but it is the only member of N which does so. This sentence is called a conditional equation, or simply equation, and it has a one-element solution set. In (i), it can be verified that -3 is the only replacement of x which satisfies the equation. However, -3 is not an element of W, the stated domain. Since no element of the given domain is a solution of (i), this equation has no root. Therefore, the equation is said to have the null set as a solution set. Alternately stated, the null set is the solution set. Braces are used around all solution sets except the null set. Here, the usual null set symbol, $\emptyset$, is used without braces. Therefore, the solution sets of (g), (h), and (i) are

R, {2}, and ∅, respectively. Please note again that a solution set may have an infinite number of elements as in (g), one element as in (h), or no elements as in (i).

In the exercises which follow, the student is asked to determine whether a number of sentences are conditional equations or identities. It is an easy task to verify that an equation is an identity. Since an identity must be true for all values of the variable in a given domain, it is only necessary to find one value of the variable for which the sentence is false. At least three or four widely varying values of the variable should be tried to verify that an equation is an identity. If the sentence is true for these values, one can be reasonably sure that it is an identity. Algebraic methods are generally necessary to positively verify that a given sentence is an identity. If an open sentence is not an identity, then it must be a conditional equation. Remember, a sentence can be a conditional equation even though it has no root.

Also, in the exercises which follow, the student is asked to find the solution sets of a number of equations by inspection. When one finds the solution set of an equation, he has **solved** the equation. More systematic methods of solving these equations, as well as others much more complicated than these, will be taught in a course of algebra.

6.3. Exercises

State whether each of the following equations are true or false with the given replacement value of the variable:

1. $4x + 1 = 9$, $x = 2$
2. $4(x + 1) = 8$, $x = 1$
3. $3y - 2 = 7$, $y = 6$
4. $5z = 3z - 2$, $z = 1$
5. $9w - 4 = 3w + 2$, $w = 1$
6. $3(x - 2) = 4(x + 1)$, $x = -10$
7. $6y - 1 = 9$, $y = \dfrac{5}{3}$
8. $2z + 5 = 5z + 9$, $z = \dfrac{-4}{3}$

Mathematical Sentences

Determine, by inspection, whether each of the following are conditional equations, or identities:

9. $3z - 6 = 3(z - 2)$

10. $2y + 3 = 6$

11. $2x - 4 = 2x + 1$

12. $\dfrac{4x - 2}{6} = \dfrac{1 - 2x}{3}$

13. $y + 1 = \dfrac{3y - 2}{4} + \dfrac{3y + 18}{12}$

14. $7z - 1 = 0$

15. $3 = 8w$

16. $\dfrac{1}{2} = \dfrac{1}{3} + \dfrac{1}{x}$

Find the solution set, by inspection, for each of the following sentences whose domain is the set of real numbers:

17. $5x = 30$

18. $-4x = 8$

19. $8y = -24$

20. $\dfrac{y}{2} = 4$

21. $\dfrac{1}{3}z = -6$

22. $\dfrac{-1}{4}w = 3$

23. $\dfrac{-1}{2}y = \dfrac{-1}{3}$

24. $2x - 1 = 7$

25. $3x = 6 - x$

26. $y = 9 + 4y$

27. $4z - 1 = 4z + 3$

28. $\dfrac{x + 8}{2} = 4 + \dfrac{x}{2}$

29. $5w - 15 = 5(w - 3)$

30. $8 - 2x = 2x - 8$

31. $\dfrac{3}{x} = 0$

32. $\dfrac{4x}{9} = 0$

Find the solution set from the stated domains for each of the following sentences:

33. $3x - 1 = 7$, $\left\{-3, 0, \dfrac{8}{3}, 4\right\}$

34. $x^2 - 2x = 3$, $\{-2, -1, 1, 3, 6\}$
35. $4y + 3 = -1$, $\{-2, 0, 1, 3\}$
36. $6(z + 3) = 6z + 3$, R

37. $8y - 9 = \dfrac{3y - 9}{3}$, $\left\{\dfrac{-6}{7}, 0, \dfrac{3}{7}, \dfrac{6}{7}\right\}$

38. $x^3 + x^2 + x + 1 = 0$, N

39. $0 \cdot z = 0$, $\left\{-8, -3, 0, \dfrac{1}{2}, 7\right\}$

40. $x^2 = x + 1$, R

6.4. Applications of Equations

The previous section was mainly concerned with introducing vocabulary which is generally used when dealing with mathematical sentences. This section shall provide some practice in translating grammatical sentences dealing with mathematical situations into their symbolic equivalents; that is, learning how to write equations. As mentioned earlier, methods for solving these equations shall be deferred to a course in algebra.

When writing algebraic expressions, it was pointed out that there are several ways to write most symbols of mathematics. This is also true of the equals sign. It can be expressed by "is," "is the same as," "is equivalent to," or simply, "is equal to." Consider these examples:

(a) Seven and two is nine.
(b) Eight less three is equivalent to five.
(c) Five is the same as three more than two.

Example (a) can be translated into the sentence $7 + 2 = 9$, example (b) into $8 - 3 = 5$, and example (c) into $5 = 3 + 2$.

Whenever a quantity is unknown, it is generally stated as "some" quantity, a "certain" age, or a "given" number. In these cases, the unknown quantity can be represented by some chosen letter, being sure to specifically write down what the letter actually represents. The student generally does not see the usefulness of this recommendation until the problems become more difficult. Consider these sentences:

(d) Nine is three less than a certain number.
(e) Five times a given quantity decreased by six is the same as twice the given quantity.
(f) The sum of Daryl's and Karen's ages is 48 years. Karen's age exceeds Daryl's by 2 years.

Mathematical Sentences

In (d), let x be the certain number. Then

$$9 = x - 3 \quad \text{is the equivalent equation}$$

In (e), let n be the given quantity. Then

$$5n \quad \text{is five times the given quantity}$$

and

$$2n \quad \text{is two times the same quantity}$$

So that

$$5n - 6 = 2n \quad \text{is the desired equation}$$

Example (f) can be written in two ways. If

$$d \quad \text{is Daryl's age in years}$$

Then

$$d + 2 \quad \text{represents Karen's age in years}$$

and therefore,

$$d + (d + 2) = 48 \quad \text{is the proper equation}$$

However, if

$$k \quad \text{represents Karen's age in years}$$

Then

$$k - 2 \quad \text{is Daryl's age in years}$$

and therefore,

$$k + (k - 2) = 48 \quad \text{becomes the correct equation}$$

Quite frequently, problems are posed which require information about more than one quantity. Each desired quantity in the problem could be represented by a variable, but this would make the resulting equation difficult or impossible to solve. The following examples illustrate how this kind of situation can be avoided:

(g) A rope 49 inches long is to be cut into two pieces. One piece is 3 inches longer than the remaining length.

(h) In a given class of 45 students, there are twice as many boys as girls.

(i) Duke is now three times as old as his son and in 12 years he will only be twice as old.

In (g), let x represent the length of the shorter piece. Then

$$x + 3 \quad \text{is the length of the longer piece}$$

Since the sum of the lengths is the whole,

$$x + (x + 3) = 49 \quad \text{is the desired equation}$$

Whenever possible, the student should draw a simple sketch to help him visualize the situation. The sketch below represents the problem in example (g):

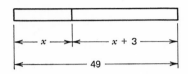

In (h), let x represent the number of girls in the class. Then

$$2x \quad \text{represents the number of boys in the class}$$

Since the sum of the boys and girls represent the total class, then the equation

$$2x + x = 45 \quad \text{represents the situation in the problem.}$$

In (i), let y represent the present age of Duke's son, and let $3y$ represent the present age of Duke. Then

$$y + 12 \quad \text{represents the age of Duke's son in 12 years}$$

and

$$3y + 12 \quad \text{is the age of Duke in 12 years}$$

In 12 years, Duke is then only twice as old as his son, so that

$$3y + 12 \quad \text{is twice as great as } y + 12$$

or

$$3y + 12 = 2(y + 12) \quad \text{is the desired equation}$$

Each of the unknown quantities above could have been represented in a different way. For example, in (h), x could have represented the number of boys and $\frac{1}{2}x$, or $\frac{x}{2}$, could represent the number of girls. The original choice was made because it did not require the use of fractions. This is generally a good rule to follow, because it is much easier to deal with equations without fractions.

Mathematical Sentences

The following problem is frequently encountered with numerous variations:

(j) Jim had a total of 12 coins which consisted of nickels and dimes. If the total value is 80 cents, write an equation in terms of the number of nickels that Jim had.

Solution: Since the number of nickels are desired, let

$$x \quad \text{represent the number of nickels}$$

There are 12 coins in all, so that the remainder will be dimes, or specifically, let

$$12 - x \quad \text{represent the number of dimes}$$

The value of x nickels is $5x$ cents, and the value of $12 - x$ dimes is $10(12 - x)$ cents. The value of the 12 coins is the sum of these, as well as 80 cents, so that

$$5x + 10(12 - x) = 80 \quad \text{is the desired equation}$$

It is useful to recognize that whenever the sum of two quantities is involved, both can be represented in terms of one variable. For example, assume that the total weight of Gary and Terrie is 265 pounds. If x represents Gary's weight in pounds, then $265 - x$, the difference, represents Terrie's weight in pounds. Again, if the total age of Bob and Sandy is 54 years, and x represents Bob's age in years, then the difference, or $54 - x$, represents Sandy's age in years.

Example (g) also involves the sum of two quantities, and the equation can be written using the same technique. Let x represent the length of

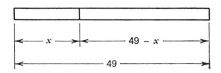

the smaller piece of rope and $49 - x$ the length of the larger piece of rope. Since the larger piece is 3 inches longer than the smaller, then $x + 3$ also represents the length of the larger piece of rope, so that $x + 3$ and $49 - x$ both represent the length of the same piece of rope. Therefore,

224 Pre-Algebra

$x + 3 = 49 - x$ is the desired equation. Although this is a different equation from that on p. 222, both yield the correct solution of 23 and 26 inches for the two lengths.

6.4. Exercises

Translate each of the following sentences into an equivalent equation. Use one variable for each problem and identify it accurately. Write the equation in terms of this one variable. Do not solve the resulting equation:

1. Four more than a certain number is eighteen. Find the number.
2. Thirteen is eight more than a certain number. Find the number.
3. A given number less four is nine. Find the number.
4. Six is some given quantity less three. Find the quantity.
5. Twelve is four less than a certain number. Find the number.
6. Some unknown quantity less than eight is the same as fourteen. Find the quantity.
7. Six more than a certain quantity is equivalent to the same quantity less than twelve. Find the quantity.
8. Nine less a given number is the same as three more than twice the given number. Find the number.
9. The total of Sandy's and Sherry's ages is 46. Sandy is four years older than Sherry. Write the equation in terms of Sandy's age.
10. Bonnie is twice as old as Jane and the total of their ages is 141. Write the equation in terms of Jane's age.
11. A board 18 feet long, is cut into two pieces. The longer piece exceeds the shorter by 4 feet. Write the equation in terms of the shorter piece.
12. Marcy has two more nickels than pennies. Write the equation in terms of the number of pennies if the total value of all her coins is 46 cents.
13. The sum of two consecutive natural numbers is 63. Write an equation in terms of the smaller number.
14. The sum of two consecutive odd integers is -24. Write an equation in terms of the smaller integer.
15. The sum of two consecutive even integers is 46. Write the equation in terms of the larger integer.

16. The sum of three consecutive integers is -39. Write the equation in terms of the smallest integer.
17. The sum of three consecutive odd integers is 57. Write the equation in terms of the middle integer.
18. The sum of three consecutive even integers is 90. Write the equation in terms of the largest integer.
19. The perimeter of a square is 44 inches. Write the equation in terms of one side of the square.
20. A rectangle has a perimeter of 92 inches and the length exceeds the width by 2 inches. Write the equation in terms of the width.
21. A log, which is 75 feet long, is cut into two pieces such that one piece is 2 feet more than three times the length of the shorter piece. Write the equation in terms of the shorter piece.
22. Three times an integer less twice the previous consecutive integer is 8. Write the equation in terms of the smaller integer.
23. Twice an integer more than five times the next larger consecutive integer is 68. Write the equation in terms of the smaller integer.
24. If one greater than twice a given number is tripled, the result is 105. Find the given number.
25. A triangle, whose perimeter is 85 inches, has two sides of equal length and a third side which is 4 inches longer than the others. Write the equation in terms of one of the two equal sides.
26. A triangle whose perimeter is 54 inches has sides which are consecutive integers. Write the equation in terms of the shortest side.
27. The second side of a given triangle, whose perimeter is 46 inches, is 4 inches longer than the first. The third side is 3 inches less than twice the length of the first side. Write the equation in terms of the first side.
28. Bob's total income is $720.00 per month, and he spends eight times as much as he saves. Write the equation in terms of the amount of money that he saves.
29. Sidney has $2.90 in his pocket, consisting of dimes and quarters, and there are 14 coins in all. Write the equation in terms of the number of quarters he has.
30. Four consecutive integers are such that the sum of the first and last is as much as the sum of the second and third. Write the equation in terms of the smallest integer.

31. Norm drives a given distance averaging 40 miles per hour (mph) and returns (the same distance) averaging 60 mph. The return trip requires 2 hours less time. Write the equation in terms of the number of hours that it takes on the return trip.
32. Irene is three times as old as Sherry. In 15 years, she will only be twice as old. Write the equation in terms of Sherry's age.

6.5. Inequalities

A second kind of mathematical sentence is the inequality. An inequality can be defined as two algebraic expressions separated by inequality symbols. The following sentences are examples of inequalities:

(a) $7 + 4 > 8$ (b) $9 + 6 < 12$ (c) $x + 3 > 6$

Observe that each of these examples contains two algebraic expressions separated by a "less than" or "greater than" symbol. The inequality in (a) is always true, while the inequality in (b) is always false. The inequality in (c) depends upon the replacement value of x as to whether it is true or false. That is, when $x = 2$, the open sentence is false, and it is true when $x = 4$. The open sentence in (c) is called a **conditional inequality.** Inequalities which are not open sentences, such as examples (a) and (b) have no purpose here and will not be discussed any further. Consider these additional examples of inequalities:

(d) $9 < 5 - x$ (e) $2x - 3 > x + 7$ (f) $2x + 7 > 2x - 3$

The student can verify that (d) is true for all values of x less than -4 and (e) is true for all values of x greater than 10. Again, these kinds of sentences which can be true or false depending upon the proper choice of the variable are called conditional inequalities. Observe that the inequality in (f) is always true, for all values of x. This kind of statement is called an **absolute inequality.** Whether or not a sentence is an absolute inequality can be determined in much the same way that an equation was checked to see if it was not an identity. That is, replace the variable with three or four widely differing values. If these make the statement true, then it is probably an absolute inequality, although not necessarily so. Generally, a formal algebraic method is necessary to positively identify an absolute inequality.

Mathematical Sentences

The solutions of inequalities are usually treated differently from the solutions of equations. The following examples will illustrate why this is necessary:

(g) $x + 1 < 2$ (h) $y + 1 < 3$ (i) $z > 0$

If the domain of each of these inequalities is N, the set of natural numbers, then (g) has no solution, (h) has one solution, namely the element 1, and (i) is true for all N. These can be written as ϕ, $\{1\}$, and N, much the same way as if they were solutions of equations.

Now consider the domain of (g) to be the set of real numbers. The solution set of (g) then becomes the set of all real numbers which are less than 1. This can easily be represented with the aid of a number line.

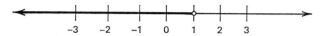

Here, the solution set is indicated by the darkened or shaded portion of the number line. The circle around the numeral 1 indicates that 1 is not an element of the solution set. This then, is a graphical representation of the solution set of (g). Now consider the symbolic representation of this solution set. The expression $x < 1$ is true when x is replaced by a number or some numbers which are less than 1. What is desired is some expression which indicates that the complete solution is the set of **all** real numbers which are less than 1. A new notation must be devised in order to write a proper symbolic representation of solution sets which are similar to (g).

The following notation is used, which the student will find to be quite simple after some practice:

$$\{x \mid x < 1, x \in R\}$$

The braces, { }, are familiar, they refer to "the set of." The first x means "all x." The vertical line reads, "such that," and the symbol, $\in$, means "is an element of," or "belongs to." The complete expression is read: "The set of all x such that x is less than 1 and x is an element of the real numbers." It would be incomplete and perhaps misleading to state a solution set of this type in any shorter way. This type of notation, used to specify a solution set, is called **set builder notation.** It is presently used in most modern books of mathematics. Some additional examples of set builder notation and their graphical equivalents will perhaps clarify this new notation:

(j) $\{x \mid x < 3, x \in R\}$

(k) $\{x \mid x > -2, x \in R\}$

(l) $\{x \mid -2 < x < 3, x \in R\}$

Example (j) is read: "The set of all x such that x is less than 3 and x is a real number." Example (k) is read: "The set of all x such that x is greater than -2 and x is a real number." Example (l) is read: "The set of all x such that x is between -2 and 3 and x is a real number." An alternate way of stating example (j), as well as the others with corresponding changes, would be: "The set of all real numbers less than 3."

These few examples should illustrate that inequalities sometimes require a more complete method of representing their solution sets. When the domain of examples (g), (h), and (i) was restricted to the set of natural numbers, their solution sets could be expressed in the same way as solutions of equations. However, when the domain of (g) was changed to the set of real numbers, the set builder notation was necessary for accuracy and clarity.

Quite frequently problems arise in which two quantities may be equal, or possibly unequal to each other. These statements require the use of the "less than or equal to," or the "greater than or equal to" inequalities which were discussed in Section 1.7, p. 33. The following statements illustrate these kinds of inequalities:

(m) If Ken gained 15 pounds, he would still weigh no more than 130 pounds. Write an expression for Ken's weight now.

(n) If Stewart Manufacturing Company costs are $3 per item, then what would be their minimum amount of profit per item, if the selling price is no less than $6.13 each?

In (m), letting w represent Ken's weight now, $w + 15 \leq 130$ is the desired inequality. In (n), letting p represent the amount of profit, then $p + 3 \geq 6.13$ would be the proper inequality.

No attempt has been made to systematically solve any of the inequalities in this section. Systematic methods of solving these inequalities will be included in a course of algebra.

The subject of inequalities has only recently become recognized as a topic of importance. Forty years ago, this material was excluded from the first year of algebra, and only briefly included in the second and third years. Today, a good first year algebra text will include a rather complete discussion of inequalities with similar treatment in the following years.

Actually, very few numerical measures are equal in the physical world. Two people seldom weigh exactly as much as each other, or are they as tall as each other. Two manufacturing companies seldom have the same output, or the same profit, or the same costs. In general, it is rare to compare two of anything in some way and find that they are equal. One will generally be less than or greater than the other. The computer, which has become an integral part of modern life, uses inequalities in many ways. For these reasons, and many others, inequalities have become an important topic in mathematics.

6.5. Exercises

State whether each of the following inequalities are true or false with the given replacement value of the variable:

1. $2x - 3 > 8$, $x = 6$
2. $3 - 4x > 11$, $x = -2$
3. $2x - 8 < 4 - x$, $x = -4$
4. $9 - 3x > 7 - 3x$, $x = 0$
5. $3(8 - x) > 4(2x - 3)$, $x = 3$
6. $\dfrac{7 - 2x}{3} < \dfrac{x - 4}{2}$, $x = 3$
7. $x^2 + 2x > x^2 - 2x$, $x = -4$
8. $-x^3 - 3x + 1 > 2x^3 - 13x + 1$, $x = -2$

Pre-Algebra

Determine whether each of the following are conditional inequalities or absolute inequalities:

9. $4x - 8 > 4(x - 8)$
10. $3x - 7 < 9$
11. $x - 2 < -2 + x$
12. $3(x - 2) > 3x - 6$
13. $2 - 7x < 8 + 3x$
14. $5x - 3 + (8 - 6x) > 4 - x$
15. $\dfrac{2}{x} < \dfrac{3}{x}$
16. $x^2 + x + 1 > 0$

Find the solution set by inspection of each of the following conditional inequalities. Graph each solution on a number line as well as stating each in set builder notation. Assume that the domain of each is the set of real numbers:

17. $x - 3 > 0$
18. $2x < 8$
19. $3x > 7$
20. $x + 1 < 5$
21. $2x + 1 < 13$
22. $\dfrac{x}{8} < 0$
23. $-4x < 12$
24. $\dfrac{1}{3}x > -2$
25. $7x > \dfrac{-1}{2}$
26. $2x - 1 > 2x + 3$
27. $4x + 8 > 0$
28. $3x - 6 > 3(x - 2)$

Write an inequality for each of the following sentences in terms of the variable specifically stated in the problem. Do not attempt to solve the resulting inequality:

29. When a given number is doubled, the result is greater than 26. Find the number.
30. Seven is less than two more than a given number which had been doubled. Find the number.
31. Two-thirds of a certain number is less than nine less one-half the given number.
32. If three less than a given number is doubled, the result is greater than the given number less five tripled.
33. The sum of two consecutive even numbers is less than 87. Let x represent the smaller number.

34. An English instructor gives 3 points for reading a book and 5 points for writing a composition. In order to pass the course, a student must have at least 42 points. Express this as an inequality letting b represent the number of books read and c represent the number of compositions written.
35. At a certain stand, profits on hamburgers are 8 cents and profits on cokes are 3 cents each. Letting h represent the number of hamburgers sold each day, and c represent the number of cokes sold each day, write an expression stating that the gross profits should exceed $46 per day.
36. A shoe store sells boots at $8 and $13 each. The store sells 12 more at $8 than they sell at $13. If the store must gross at least $300 on boots, write an expression letting x represent the number of $13 boots.

6.6. Summary

In this chapter an attempt has been made to show the student that there is a strong similarity between grammatical and mathematical sentences. Just as the grammatical sentence consists of phrases, so does the mathematical sentence, which is called an algebraic expression. In order for the written sentence to be grammatically correct, it must include a verb. This is also true for the mathematical sentence, and here the relation takes the place of the verb. The equality and inequality are the two principal relations studied. The main emphasis here was to become exposed to writing mathematical sentences, both equations and inequalities. The formal solutions of these sentences were not discussed so that they could be properly studied in an algebra course later. Hopefully, the student will be better prepared for their solutions by his exposure to the topic here.

Also briefly introduced here was the set builder notation, a device extensively used in mathematics to properly represent symbolically the solutions of inequalities.

Much of what has been presented here may not be seen again by many students reviewing this material. The "variables" involved are the preferences of your instructor and the time available in the algebra course which you shall take. Those students continuing beyond the first

232   Pre-Algebra

year must surely be exposed to mathematics in sentence form. The student who continues in mathematics will be severely handicapped if he cannot translate mathematically oriented material into symbolic form.

6.6. Review Exercises

Translate, exactly as stated, each of the following statements into algebraic expressions. Do not perform the indicated operations:

1. Seven more than four
2. Nine less than four
3. The double negative of three
4. The absolute value of negative two
5. Twice a given number less eight
6. The difference of the squares of nine and four
7. The square of the difference of nine and four
8. Twelve less the cube of a given number
9. The sum of twice a given number and seven less three times a second number
10. The sum of the squares of two consecutive numbers in terms of the smaller number
11. The negative of the square of a certain number
12. The square of the quotient of twice a given number and three times that number less five

Find a suitable algebraic expression which most nearly expresses each of the following statements:

13. If x represents Mark's age, express the age of Lori, who is 2 years younger.
14. If x represents Taylor's age, express the age of Gary, who is six times as old.
15. A picture is 3 inches wider than its height.
    (a) If w represents the width, express the height in terms of its width.
    (b) If h represents the height, express the width in terms of its height.
16. Mike is 2 years younger than Rosie and 3 years older than Sandy.

(a) Express their ages in terms of m, Mike's age.
(b) Express their ages in terms of r, Rosie's age.
(c) Express their ages in terms of s, Sandy's age.
17. Bernice is 7 years younger than four times Carol's age.
    (a) Express Bernice's age in terms of c, Carol's age.
    (b) Express Carol's age in terms of b, Bernice's age.
18. The Schall Manufacturing Company made s dollars in total sales and had expenses of $17,000. Write an expression for the gross profit that the company made in terms of x.
19. In a two-digit number, the tens digit is three times the units digit. Write an expression for the number in terms of u, the units digit.
20. Emma is 16 years younger than twice Irene's age. Express Emma's age 8 years from now in terms of x, Irene's age now.

State whether each of the following sentences are true or false with the given replacement value of the variables:

21. $3x - 4 = 8$, $x = 4$

22. $4x + 1 < 8$, $x = -2$

23. $2x - 3 > 4x + 5$, $x = -4$

24. $6y - 7 = 9 - 3y$, $y = \dfrac{16}{9}$

25. $2(z - 3) = 3z + 5$, $z = -8$

26. $\dfrac{2y}{3} < 4$, $y = 5$

27. $2x^3 > 3x^2 + x + 1$, $x = -2$

28. $\dfrac{3 - 2x}{3} = \dfrac{4x - 1}{4}$, $x = \dfrac{3}{4}$

Determine by inspection, the type of each of the following sentences—conditional inequality, conditional equation, identity, or absolute inequality:

29. $3x - 1 > 8$

30. $3y - 2 = 5$

31. $3z + 7 = 7 + 3z$

32. $\dfrac{4z + 8}{4} > z + 1$

33. $3(x - 2) > 3(x - 1)$

34. $\dfrac{3}{x} + \dfrac{2}{x} = 7$

35. $7(x + 3) = 7x + 21$

36. $1 - 2x - x^2 < 0$

Find the solution set by inspection, for each of the following sentences whose domain is the set $\{-6, -5, -4, \ldots, 7, 8, 9, 10\}$. Write your solution as a set:

37. $3x = 12$
38. $x < 4$
39. $2x = -6$
40. $\dfrac{x}{2} > 3$

41. $2x + 1 = 13$
42. $-4x = -16$
43. $5x > x + 12$
44. $4x - 8 > 4(x - 2)$

Find the solution set by inspection, for each of the following inequalities whose domain is the set of real numbers. Write your solution in set builder notation as well as graphing on a number line:

45. $x + 1 > -3$
46. $2x - 3 < 7$

47. $x > -4$ and $2x < 6$
48. $4x + 3 > 4x - 1$

Translate each of the following sentences into an equivalent equation or inequality—do not solve:

49. Two more than three times a given quantity is nine. Find the number.
50. Seven less than twice a certain number is the same as that number less four.
51. The sum of two consecutive integers is -31. Write an equation in terms of the larger integer.
52. Twice a given number less nine is greater than seventeen. Find the number.
53. A doctor told a woman that she must lose 2 pounds a month for a period of 5 months, but her weight at the end of that period must not exceed 120 pounds. Write an expression indicating this relationship letting x represent her weight now.
54. The perimeter of a certain rectangle is less than 56 inches. The width is 3 inches less than the length. Write an expression showing this relationship letting w represent the one variable used.

7
Numeral Systems

7.1. Base Five Numerals

The basic types of numeral systems were discussed in Chapter 1: simple grouping and positional systems. In this chapter, there will be an opportunity to explore positional numeral systems other than the Hindu–Arabic system. Those students who have not been exposed to these systems before shall be pleasantly surprised at their usefulness and importance.

Although they are different types of numeral systems, the Egyptian, Roman, and Hindu–Arabic systems, discussed in Chapter 1, all have one property in common. The symbols in the grouping systems basically differed by a factor of 10, and the place values in the positional system did the same. This number, which a positional system is built around, is called its **base.** The Hindu–Arabic numeral system is a base 10 system since each place or position is a multiple of a power of ten. It is also called a **decimal** system after the Latin prefix *decem*, which means ten. Perhaps the reader has already assumed that ten was used as the base because of

238   Pre-Algebra

the number of digits either on both hands or both feet. Actually, historians have conjectured that primitive man did structure some of his numeral systems in this natural way. There were also some ancient numeral systems which were not based upon powers of ten.

There is ample evidence that the Babylonians, sometime between 2000 BC and 3000 BC, used a numeral system with two symbols: | (which has a value of 1), and < (which has a value of 10). A simple grouping system was used to form numbers which were less than 60. For example:

(a) $<<<|\,| = 32$
(b) $<<|\,|\,|\,|\,|\,|\,| = 27$
(c) $<<<<|\,|\,|\,|\,|\,| = 46$

However, for numbers larger than 60, a positional system was used. Each position in the numeral was based upon successively larger powers of 60. That is, the Babylonians used a base 60 positional numeral system for numbers larger than 60. A few examples should clarify this idea:

(d) $<|\genfrac{}{}{0pt}{}{<<}{<<}<|\,|\,| = 11 \cdot 60 + 53$
$= 713$

(e) $|\,|\genfrac{}{}{0pt}{}{<}{<}|<|\,| = 2 \cdot 60^2 + 21 \cdot 60 + 12$
$= 7{,}200 + 1{,}260 + 12$
$= 8{,}472$

(f) $\genfrac{}{}{0pt}{}{<}{<}|\,|\,|\genfrac{}{}{0pt}{}{<}{<}<|\genfrac{}{}{0pt}{}{<<}{<<}|\,|\,|\,|\,|\,|<|\,| = 23 \cdot 60^3 + 31 \cdot 60^2 + 46 \cdot 60 + 12$
$= 4{,}968{,}000 + 111{,}600 + 2{,}760 + 12$
$= 5{,}082{,}372$

It was sometimes difficult to separate the digits in the Babylonian system, and a zero was very much needed, which however, had not been invented yet. Regardless of these shortcomings, the system was surprisingly modern considering that the Hindu–Arabic system was not introduced to the Western world for another 3,000 years. There is some conjecture that the base of 60 was used because it is an exact factor of 360, which was then thought to be the number of days in a year. At any rate, the influence of the Babylonian system is still felt today in timekeeping, angle measurement, and navigation, which are all based upon multiples of 60.

Another unusual base uncovered in the sixteenth century was that used by the Mayans in Yucatan. They used a positional system with a base of 20, with the exception of the second digit from the right. This digit represented multiples of 18. This system was also felt to be based upon the 360 day year which the Mayans used.

Historians have uncovered other systems using base 3 and base 5, demonstrating that numeral systems using a base 10 should not be considered sacred. For these reasons, it should not seem unnatural to develop numeral systems using bases other than 10.

The base 10 system perhaps originated when man placed objects in a pile, counting these off against his fingers. Mathematicians refer to this principle as using a **1-to-1 correspondence.** That is, man matched one object against one finger until no objects remained unmatched. The number of fingers, which were matched against objects, were the same as the number of objects. Man then gave this number of fingers a name and later represented this name by a numeral. Using this system, twelve objects would be arranged in one pile and two would be left over. Incidentally, the word twelve is derived from two words, *twe lef,* meaning two left over. It should be easy to visualize that this concept can be extended to the development of one "superpile," which contains ten piles. It shall be convenient to identify this "super-pile" by the familiar notation, 10^2. A "super-super-pile" can then be denoted by 10^3, and succeeding piles by 10^4, 10^5, etc.

Now visualize that the above objects are grouped into piles of five, so that the twelve objects are in two groups of five with two left over. This can be written as $2 \cdot 5 + 2$. Similarly, sixteen objects can be placed in one pile of ten with six left over, or three piles of five with one left over. The former is written as $1 \cdot 10 + 6$, and the latter as $3 \cdot 5 + 1$. Writing these in positional notation, $1 \cdot 10 + 6$ becomes 16, while $3 \cdot 5 + 1$ becomes 31. Unfortunately, the 31 is indistinguishable from thirty-one ($3 \cdot 10 + 1$), so it must be identified in some additional way to avoid confusion. This is usually done by writing a **subscript** after the numeral which represents the base used. Subscripts are written slightly below the numeral and are not to be confused with exponents which are written slightly above the numeral. Therefore, 31 base 5 is written as 31_5. To avoid any additional confusion, 31_5 is read "three one base five" and **not** "thirty-one base 5," which is a meaningless expression. The term "thirty-one" is reserved for the quantity of 3 piles of 10 with 1 left over. It shall be understood that a numeral without a subscript has a base of 10.

Continuing then, four piles of five with 3 left over can be written as 43_5, which is equivalent to $4 \cdot 5 + 3$ or 23. Five piles of five become one super-pile as before, and shall be designated by the notation 5^2. One super-pile with one left over can be written as $1 \cdot 5^2 + 0 \cdot 5 + 1$. The term $0 \cdot 5$ is included to indicate that there are no piles of five. The expanded notation $1 \cdot 5^2 + 0 \cdot 5 + 1$ can then be written as 101_5 in base five positional notation. By dispensing with the invented expressions "pile" and "super-pile," it shall soon be obvious that any quantity can either be expressed in expanded notation or positional notation using powers of five or a base of five. The following examples illustrate this last statement:

(g) $\quad 3 \cdot 5 + 2 = 32_5$
(h) $\quad 4 \cdot 5^2 + 3 \cdot 5 + 1 = 431_5$
(i) $\quad 2 \cdot 5^3 + 4 \cdot 5^2 + 4 = 2404_5$

In (g), $3 \cdot 5 + 2 = 17$, so that $32_5 = 17$. In (h),

$$4 \cdot 5^2 + 3 \cdot 5 + 1 = 100 + 15 + 1 = 116,$$

so that $431_5 = 116$. In (i), $2 \cdot 5^3 + 4 \cdot 5^2 + 4 = 250 + 100 + 4 = 354$, so that $2404_5 = 354$.

Examples (g), (h), and (i), along with the preceding discussion, should convince the reader that any quantity can be placed into piles and super-piles of five. Restating this, any quantity can be expressed in base five notation. The base ten equivalent of any numeral written in base five positional notation can be found by first writing it in expanded notation. The sum of multiples of powers of five can then be calculated and written as a base ten numeral. Observe the following examples:

(j) $\quad 34_5 = 3 \cdot 5 + 4$
$\qquad\quad = 15 + 4$
$\qquad\quad = 19$

(k) $324_5 = 3 \cdot 5^2 + 2 \cdot 5 + 4$
$= 75 + 10 + 4$
$= 89$

(l) $1423_5 = 1 \cdot 5^3 + 4 \cdot 5^2 + 2 \cdot 5 + 3$
$= 125 + 100 + 10 + 3$
$= 238$

(m) $10342_5 = 1 \cdot 5^4 + 0 \cdot 5^3 + 3 \cdot 5^2 + 4 \cdot 5 + 2$
$= 625 + 0 + 75 + 20 + 2$
$= 722$

The task of converting base ten numerals into their base five equivalents can be more difficult. Remembering that since base five numerals occur in multiples of 5, 5^2, 5^3, etc., then multiples of 5, 25, 125, etc., respectively, can be subtracted from a given base ten numeral. Consider the numeral 27. It does not include a quantity of 5^3 or 125, but does include one 5^2 or 25. Therefore, $27 = 25 + 2$ or $1 \cdot 5^2 + 0 \cdot 5 + 2$, which can be written as 102_5. Consider as a second example the numeral 326. Upon checking, it is found to be between 250 ($2 \cdot 5^3$) and 375 ($3 \cdot 5^3$). The remainder of 76, found after subtracting 250 from 326, contains $3 \cdot 5^2$ or 75. Subtracting 75 from 76 leaves a remainder of 1. Therefore, $326 = 2 \cdot 5^3 + 3 \cdot 5^2 + 0 \cdot 5 + 1$ or 2301_5. This method can be trying to those who are somewhat slow with numbers. Fortunately, a simpler algorithm produces the same result with much less effort. Most students will be content to find that the algorithm does work so that no time shall be lost in trying to explain why, although it is based on the discussion above.

To illustrate this new algorithm, consider the numeral 27 again, which had just been converted to 102_5. The method consists of dividing by the base, in this case five, until the quotient is zero. After each division, indicate the remainder of the result at right. Reading the remainders from top to bottom produces the desired numerals in base five. The calculations below show the details of this conversion:

```
      0
   5)1      —The quotient of   5 by 1  with a remainder  1
   5)5      —The quotient of   5 by 5  with a remainder  0
  5)27     —The quotient of  27 by 5  with a remainder  2
```

—The desired numeral in base five is 102_5

242 Pre-Algebra

Consider now a second example, the numeral 326, which was found to be equivalent to 2301_5. The algorithm is again displayed below, letting R be the abbreviation for remainder.

$$
\begin{array}{ll}
0 & -R \quad 2 \\
5\overline{)2} & -R \quad 3 \\
5\overline{)13} & -R \quad 0 \\
5\overline{)65} & -R \quad 1 \\
5\overline{)326} &
\end{array}
$$

—The desired numeral in base five is 2301_5

A final example should make the algorithm clear.

(n) Convert 743 to a base five numeral.

$$
\begin{array}{ll}
0 & -R \quad 1 \\
5\overline{)1} & -R \quad 0 \\
5\overline{)5} & -R \quad 4 \\
5\overline{)29} & -R \quad 3 \\
5\overline{)148} & -R \quad 3 \\
5\overline{)743} &
\end{array}
$$

—The desired numeral is $10{,}433_5$

It is sometimes helpful to insert commas every three digits for ease of reading. It would be interesting to now convert $10{,}433_5$ back into a base ten numeral in order to confirm that the result is correct. This can be done by first writing the numeral in expanded notation and then finding the sum of the products.

$$
\begin{aligned}
10{,}433_5 &= 1 \cdot 5^4 + 0 \cdot 5^3 + 4 \cdot 5^2 + 3 \cdot 5 + 3 \\
&= 625 + 0 + 100 + 15 + 3 \\
&= 743
\end{aligned}
$$

In this section a method has been developed by which any base ten numeral can be expressed as a base five numeral, and any base five numeral can be expressed in base ten notation. The exercises which follow should provide the reader with enough practice to become competent with these conversions.

One final word of caution. When making conversions, remember that the set of numerals in base five is $\{0,1,2,3,4\}$. A base five numeral cannot contain a digit as large or larger than 5 for the same reason that Hindu–Arabic numerals cannot have digits which are larger than 9. The digits

Numeral Systems

of any numeral in a positional numeral system must always be less than the base.

7.1. Exercises

Write each of the following Hindu–Arabic numerals as a Babylonian numeral:

1. 23
2. 47
3. 84
4. 144

Write each of the following Babylonian numerals as Hindu–Arabic numerals. Ample spacing has been provided between digits.

5. <<<<|||
6. <<<|||||||
7. |$\overset{<}{<}$<|||
8. |||$\overset{<}{<}$||
9. | <| <|
10. | | <| $\overset{<}{<}$||

Write each of the following sums as a numeral in base five notation:

11. $4 \cdot 5 + 1$
12. $3 \cdot 5^2 + 2 \cdot 5 + 4$
13. $4 \cdot 5^3 + 2 \cdot 5^2 + 5 + 3$
14. $3 \cdot 5^4 + 2 \cdot 5^3 + 1 \cdot 5^2 + 4 \cdot 5 + 3$
15. $2 \cdot 5^3 + 4 \cdot 5 + 3$
16. $5^3 + 4$
17. $2 \cdot 5^4 + 3 \cdot 5^2$
18. $3 \cdot 5^5$

Convert each of the following base five numerals into an equivalent Hindu–Arabic numeral:

19. 34_5
20. 13_5
21. 112_5
22. 321_5
23. 1234_5
24. 4321_5
25. 10_5
26. 204_5
27. 2030_5
28. 3100_5

Convert each of the following Hindu–Arabic numerals into an equivalent base five numeral:

29. 19
30. 15
31. 48
32. 93
33. 147
34. 396
35. 781
36. 1348

7.2. Operations with Base Five Numerals

Working with a new numeral system is very much like learning a new language. When learning a new language, all objects are first thought of in the familiar tongue, in this case English. When learning a new numeral system, the numerals are more easily understood in terms of Hindu–Arabic numerals. These are more familiar because of long usage and also the fact that each numeral has a convenient name. The expression, thirty-one, is much easier to work with than one one one base five. If, for some reason, base five numerals were used a great deal, then names could be invented for these and the identifying label "base five" could also be omitted. There is no intention of doing that here.

Before leaving base five numerals to study numeral systems with other bases, it should be interesting to the reader to perform some arithmetic operations with base five numerals. The student shall be pleased to find that these new techniques are not as difficult to learn as he may imagine.

First consider the addition problem $3_5 + 4_5$. Since $3_5 = 3$ and $4_5 = 4$, then $3_5 + 4_5 = 3 + 4 = 7$ and $7 = 12_5$ so that $3_5 + 4_5 = 12_5$. The remaining base five numerals 0_5, 1_5, and 2_5, are also equal to 0, 1, and 2, respectively, so that the sum of any two one-digit numerals can be found in the same way. A table is conveniently provided below with these sums:

——Column

+	0	1	2	3	4
0	0	1	2	3	4
1	1	2	3	4	10
2	2	3	4	10	11	← Row
3	3	4	10	11	12
4	4	10	11	12	13

Each entry in the table is a base five numeral. Horizontal entries are called rows and vertical entries are called columns, as indicated above.

The sum $3_5 + 4_5$ is the entry at the intersection of the row opposite 3 and the column opposite 4. Likewise, $4_5 + 2_5 = 11_5$, the entry in the row opposite 4 and the column opposite 2.

Observe that $3_5 + 2_5 = 2_5 + 3_5 = 10_5$. Careful examination of the table will reveal to the reader that the commutative property of addition holds for these numerals and can be shown to be generally true. The associative property for addition is also valid for base five numerals. The following example illustrates this property, however, it is difficult to prove it generally:

(a) Establish that

$$(12_5 + 34_5) + 23_5 = 12_5 + (34_5 + 23_5)$$

On the left side of the expression,

$$12_5 + 34_5 = 101_5 \quad \text{and} \quad 101_5 + 23_5 = 124_5$$

Therefore,

$$(12_5 + 34_5) + 23_5 = 124_5$$

On the right side of the expression,

$$34_5 + 23_5 = 112_5 \quad \text{and} \quad 12_5 + 112_5 = 124_5$$

Therefore,

$$12_5 + (34_5 + 23_5) = 124_5$$

Both sides of the expression are equal, establishing that the associative property holds for this particular choice of numerals.

The results of Table 7.1 could be memorized, but it is not recommended. It is far easier to add one-digit numerals in base ten and then mentally convert the results to base five. For example, since $4_5 + 4_5 = 4 + 4 = 8$, and $8 = 1 \cdot 5 + 3$, then it follows that $4_5 + 4_5 = 13_5$.

Now to examine addition involving more than one-digit numerals.

(b) Find the sum of 13_5 and 4_5.

$$\begin{aligned} 13_5 + 4_5 &= (1 \cdot 5 + 3) + 4 \\ &= 1 \cdot 5 + (3 + 2) + 2 \\ &= 2 \cdot 5 + 2 \\ &= 22_5 \end{aligned}$$

246   Pre-Algebra

The above sum can be more easily found by writing the problem vertically and then using the familiar addition algorithm. Remember that the additions are in base five.

$$\begin{array}{r} {}^1 1\ 3_5 \\ 4_5 \\ \hline 2 \end{array}$$ Since $3 + 4 = 12_5$, the 2 is "brought" down and the 1 is "carried" to the next column

$$\begin{array}{r} {}^1 1\ 3_5 \\ 4_5 \\ \hline 2\ 2_5 \end{array}$$ Finishing the addition, $1 + 1 = 2$

Some further examples should make this algorithm clear.

(c)   Find the sum $24_5 + 33_5$.

$$\begin{array}{r} {}^1 2\ 4_5 \\ 3\ 3_5 \\ \hline 2 \end{array}$$ Since $4_5 + 3_5 = 12_5$, the 2 is "brought" down and the 1 is "carried"

$$\begin{array}{r} {}^1 2\ 4_5 \\ 3\ 3_5 \\ \hline 1\ 1\ 2_5 \end{array}$$ Since $1 + 2 + 3 = 11_5$, the 11 is "brought" down

Checking this result in base ten,

$$\begin{array}{rl} 2\ 4_5 = 2 \cdot 5 + 4 & = 14 \\ 3\ 3_5 = 3 \cdot 5 + 3 & = 18 \end{array} \bigg\} 32$$
$$\overline{1\ 1\ 2_5} = 1 \cdot 5^2 + 1 \cdot 5 + 2 = 32$$

(d)   Find the sum $312_5 + 414_5$

$$\begin{array}{r} 3\ {}^1 1\ 2_5 \\ 4\ 1\ 4_5 \\ \hline 1 \end{array}$$ Since $2_5 + 4_5 = 11_5$, the 1 is "brought" down and 1 is "carried"

$$\begin{array}{r} 3\ {}^1 1\ 2_5 \\ 4\ 1\ 4_5 \\ \hline 3\ 1 \end{array}$$ Since $1 + 1 + 1 = 3$, the 3 is "brought" down.

$$\begin{array}{r} 3\ {}^1 1\ 2_5 \\ 4\ 1\ 4_5 \\ \hline 1\ 2\ 3\ 1_5 \end{array}$$ Since $3_5 + 4_5 = 12_5$, the 12 is "brought" down

Checking again in base ten,
$$3\ 1\ 2_5 = 3 \cdot 5^2 + 1 \cdot 5 + 2 \qquad = 82$$
$$4\ 1\ 4_5 = 4 \cdot 5^2 + 1 \cdot 5 + 4 \qquad = 109 \Big\} 191$$
$$\overline{1\ 2\ 3\ 1_5} = 1 \cdot 5^3 + 2 \cdot 5^2 + 3 \cdot 5 + 1 = 191$$

Multiplication can also be performed in base five. Remembering that $2_5 \cdot 3_5 = 2 \cdot 3 = 6$, and since $6 = 11_5$, then $2_5 \cdot 3_5 = 11_5$. A multiplication table of one-digit products is given below:

·	0	1	2	3	4
0	0	0	0	0	0
1	0	1	2	3	4
2	0	2	4	11	13
3	0	3	11	14	22
4	0	4	13	22	31

Again, each entry in the table is a base five numeral, and the products are found here in the same way as the sums were found in the addition table. For example, $3_5 \cdot 4_5 = 22_5$, which is found in the row opposite 3 and the column opposite 4. It is not recommended that this table be memorized. The above method of finding the products in base ten is preferred. That is, $3_5 \cdot 4_5 = 3 \cdot 4 = 12 = 2 \cdot 5 + 2 = 22_5$.

Observe from the table that $3_5 \cdot 4_5 = 4_5 \cdot 3_5 = 22_5$. It can be seen from the table that the product of any two one-digit numerals is commutative. The product of any two base five numerals is commutative, however, this will not be proven here.

Now to consider the product of a two-digit numeral by a one-digit numeral, both in base five. The product $4_5 \cdot 13_5$ will provide an example:

(e) $4_5 \cdot 13_5 = 4 \cdot (1 \cdot 5 + 3)$
$\qquad\qquad = 4 \cdot 1 \cdot 5 + 4 \cdot 3 \qquad$ Distributive property
$\qquad\qquad = 4 \cdot 5 + 2 \cdot 5 + 2 \qquad 4 \cdot 3 = 2 \cdot 5 + 2$
$\qquad\qquad = 1 \cdot 5^2 + 1 \cdot 5 + 2 \qquad 4 \cdot 5 + 2 \cdot 5 = 1 \cdot 5^2 + 1 \cdot 5$
$\qquad\qquad = 112_5$

248   Pre-Algebra

Fortunately, the multiplication algorithm for base ten is also valid with base five numerals. Again, remember that all multiplications must correspond to Table 7.2. The following steps in vertical form should help to make the algorithm clear:

$$\begin{array}{r} \overset{2}{1}\,3_5 \\ 4_5 \\ \hline 2 \end{array}$$
Since $4_5 \cdot 3_5 = 22_5$, a 2 is "brought" down and a 2 is "carried"

$$\begin{array}{r} \overset{2}{1}\,3_5 \\ 4_5 \\ \hline 1\,1\,2_5 \end{array}$$
Since $4 \cdot 1 = 4$ and $4 + 2 = 11_5$, 11 is "brought" down

This checks with the more lengthy method above. Now consider the following product of two-digit numerals: $13_5 \cdot 24_5$. The product can be found by first writing these in expanded notation and then using the distributive property.

(f) $\quad 13_5 \cdot 24_5 = (1 \cdot 5 + 3)(2 \cdot 5 + 4)$
$\qquad\qquad\quad = 1 \cdot 5(2 \cdot 5 + 4) + 3(2 \cdot 5 + 4)$
$\qquad\qquad\quad = 1 \cdot 5(2 \cdot 5) + 1 \cdot 5 \cdot 4 + 3 \cdot 2 \cdot 5 + 3 \cdot 4$
$\qquad\qquad\quad = 2 \cdot 5^2 + 4 \cdot 5 + (5 + 1) \cdot 5 + (2 \cdot 5 + 2)$
$\qquad\qquad\quad = 3 \cdot 5^2 + 5^2 + 2 \cdot 5 + 2$
$\qquad\qquad\quad = 4 \cdot 5^2 + 2 \cdot 5 + 2$
$\qquad\qquad\quad = 422_5$

This product can be checked by first converting to base ten.

$13_5 \cdot 24_5 = (1 \cdot 5 + 3)(2 \cdot 5 + 4)$
$\qquad\quad\ = 8 \cdot 14$
$\qquad\quad\ = 112$
$422_5 = 4 \cdot 5^2 + 2 \cdot 5 + 2$
$\qquad = 100 + 10 + 2$
$\qquad = 112$

Therefore, 422_5 is the correct product of 13_5 and 24_5.

Numeral Systems 249

It can be seen that the task is far easier when performed with the multiplication algorithm.

$$\begin{array}{r} \overset{2}{1}\ 3_5 \\ 2\ 4_5 \\ \hline 1\ 1\ 2 \end{array}$$ The first product, $4_5 \cdot 13_5 = 112_5$, as performed above

$$\begin{array}{r} \overset{1}{1}\ 3_5 \\ 2\ 4_5 \\ \hline 1\ 1\ 2 \\ 3\ 1 \\ \hline 4\ 2\ 2_5 \end{array}$$ The second product, $2_5 \cdot 13_5 = 31_5$, as shown

The sum agrees with the result found above. When these two steps are combined, the process is shortened considerably. A final example will illustrate the process further:

(g) Find the product of 234_5 and 32_5.

$$\begin{array}{r} \overset{1}{2}\overset{1}{3}\ 4_5 \\ 3\ 2_5 \\ \hline 1,0\ 2\ 3 \end{array}$$ The first step at left is $2_5 \cdot (234_5) = 1{,}023_5$, with the "carrying" marks as indicated

$$\begin{array}{r} \overset{2}{2}\overset{2}{3}\ 4_5 \\ 3\ 2_5 \\ \hline 1\ 0\ 2\ 3 \\ 1\ 3\ 1\ 2 \\ \hline 1\ 4{,}1\ 4\ 3_5 \end{array}$$ The second step is $3_5(234_5) = 1{,}312_5$, with the "carrying" marks again indicated.

A final check is made by conversion to base ten.

$$\begin{aligned} 234_5 &= 2 \cdot 5^2 + 3 \cdot 5 + 4 \\ 32_5 &= 3 \cdot 5 + 2 \\ \hline 14{,}143_5 &= 1 \cdot 5^4 + 4 \cdot 5^3 + 1 \cdot 5^2 + 4 \cdot 5 + 3 \\ &= 625 + 500 + 25 + 20 + 3 \\ &= 1{,}173 \end{aligned}$$

$$\begin{array}{r} 6\ 9 \\ 1\ 7 \\ \hline 4\ 8\ 3 \\ 6\ 9 \\ \hline 1{,}1\ 7\ 3 \end{array}$$

The associative property of multiplication for base five numerals is also valid, although it shall be assumed at this time. A specific example will

suffice to demonstrate this property:
(h) Verify that $(3_5 \cdot 4_5) \cdot 2_5 = 3_5 \cdot (4_5 \cdot 2_5)$. Since
$$(3_5 \cdot 4_5) \cdot 2_5 = (22_5) \cdot 2_5$$
$$= 44_5$$
and
$$3_5 \cdot (4_5 \cdot 2_5) = 3_5 \cdot (13_5)$$
$$= 44_5$$
then
$$(3_5 \cdot 4_5) \cdot 2_5 = 3_5 \cdot (4_5 \cdot 2_5)$$
and the associative property is verified for these specific numerals.

The remaining arithmetic operations of subtraction and division shall not be discussed here. The student who is so inclined should be able to perform these operations by himself with a little practice. The base ten algorithms for these operations are valid with appropriate changes.

In this section then, addition and multiplication of base five numerals have been demonstrated. Algorithms have been developed, which with the aid of some practice exercises, should become almost as familiar to work with as those in base ten. The associative and commutative properties both for addition and multiplication were shown to apply for all base five numerals.

7.2. Exercises

Perform the operations as indicated on each of the following. Check your results by converting to base ten as in examples (f) and (g):

1. $21_5 + 14_5$
2. $32_5 + 43_5$
3. $13_5 + 33_5$
4. $42_5 + 24_5$
5. $12_5 \cdot 3_5$
6. $23_5 \cdot 4_5$
7. $12_5 \cdot 21_5$
8. $23_5 \cdot 13_5$
9. $41_5 \cdot 24_5$
10. $34_5 \cdot 23_5$
11. $423_5 + 112_5$

12. $231_5 + 404_5$
13. $312_5 + 143_5$
14. $431_5 + 223_5$
15. $214 \cdot 3_5$
16. $321_5 \cdot 4_5$
17. $232_5 \cdot 12_5$
18. $421_5 \cdot 34_5$
19. $1{,}402_5 + 314_5$
20. $213_5 + 1{,}324_5$
21. $32{,}231_5 + 1{,}322_5$
22. $40{,}314_5 + 1{,}224_5$
23. $304_5 \cdot 123_5$
24. $412_5 \cdot 240_5$
25. $2{,}301_5 + 3{,}040_5 + 4{,}302_5$
26. $4{,}100_5 + 2{,}341_5 + 1{,}004_5 + 4{,}123_5$
27. $3{,}020_5 \cdot 2{,}003_5$
28. $4{,}013_5 \cdot 2{,}304_5$

Verify that the associative property holds for each of the following examples.

29. $(23_5 + 34_5) + 41_5 = 23_5 + (34_5 + 41_5)$
30. $412_5 + (231_5 + 143_5) = (412_5 + 231_5) + 143_5$
31. $(23_5 \cdot 34_5) \cdot 41_5 = 23_5 \cdot (34_5 \cdot 41_5)$
32. $14_5 \cdot (32_5 \cdot 43_5) = (14_5 \cdot 32_5) \cdot 43_5$

7.3. Numerals Other Than Base Five

It was demonstrated in Section 7.1 that numbers written with base ten numerals could easily be converted to equivalent numbers which were written as base five numbers. This simply means that any number of objects can be counted in piles of ten or piles of five. There seems to be no reason why a given number of objects could not be counted in piles of any quantity. Since the various digits in a positional numeral system are determined by successive powers of the base, different powers of 0 or 1 do not change, and therefore, these must be exluded as possible choices of a base.

252   Pre-Algebra

Consider now a numeral system using a base of three. This system is restricted to only using the digits, 0, 1, and 2. Numerals in this base use powers of 3 such as $3^2 = 9$, $3^3 = 27$, $3^4 = 81$, $3^5 = 243$, etc. The numeral 17 can be written as $9 + 6 + 2$ or $1 \cdot 3^2 + 2 \cdot 3 + 2$, which becomes the base three numeral, 122_3. The conversion to base three can also be accomplished as before with the use of the division algorithm as shown below:

$$\begin{array}{rl} 0 & -R \quad 1 \\ 3\overline{)1} & -R \quad 2 \\ 3\overline{)5} & -R \quad 2 \\ 3\overline{)17} & \end{array}$$

Here again, $17 = 122_3$ with perhaps less difficulty in obtaining these results than before. Another example is in order here:

(a)   Convert 267 to a base three numeral and check the result in base ten.

$$\begin{array}{rl} 0 & -R \quad 1 \\ 3\overline{)1} & -R \quad 0 \\ 3\overline{)3} & -R \quad 0 \\ 3\overline{)9} & -R \quad 2 \\ 3\overline{)29} & -R \quad 2 \\ 3\overline{)89} & -R \quad 0 \\ 3\overline{)267} & \end{array}$$

—The desired result is $100,220_3$

Checking,
$$\begin{aligned} 100,220_3 &= 1 \cdot 3^5 + 0 \cdot 3^4 + 0 \cdot 3^3 + 2 \cdot 3^2 + 2 \cdot 3 + 0 \\ &= 243 + 0 + 0 + 18 + 6 + 0 \\ &= 267 \end{aligned}$$

Here then is an example of a base ten numeral being converted into base three and a base three numeral being converted into base ten. Without trying to justify the statement, let it suffice to say that this can be done generally for all numbers.

Operations can also be performed with base three numerals. The following examples should be verified by the student:

(b) $21_3 + 11_3 = 102_3$
(c) $12_3 + 12_3 = 101_3$
(d) $2_3 \cdot 2_3 = 11_3$
(e) $12_3 \cdot 21_3 = 1{,}022_3$

Checking example (e) by first converting to base ten,
$$12_3 = 5 \quad \text{and} \quad 21_3 = 7$$
so that
$$5 \cdot 7 = 35$$
Also,
$$1{,}022_3 = 1 \cdot 3^3 + 0 \cdot 3^2 + 2 \cdot 3 + 2 = 35$$
which verifies the product.

The same principle as above can be used to write numerals in other bases. Consider the writing of numerals in base seven. The conversion to base seven can be easily performed with the same division algorithm used before, as seen in the following example:

(f) Convert 163 to a base seven numeral.

$$\begin{array}{r} 0 \\ 7\overline{)3} \\ 7\overline{)23} \\ 7\overline{)163} \end{array} \quad \begin{array}{l} -R \\ -R \\ -R \end{array} \quad \begin{array}{c} 3 \\ 2 \\ 2 \end{array}$$

—The desired numeral is 322_7

This result can also be checked in the usual way.
$$\begin{aligned} 322_7 &= 3 \cdot 7^2 + 2 \cdot 7 + 2 \\ &= 147 + 14 + 2 \\ &= 163 \end{aligned}$$

Here again is an example of a base ten numeral being converted into a base seven numeral and a base seven numeral written in base ten. There shall be no effort made here to prove these results generally; however, it can be done. Operations in any base less than ten can be performed with the same ease as with base five or base three. It shall be left to the student to gain familiarity with numeral systems using these other bases. Numerous exercises shall be provided at the end of this section for that purpose.

Numeral systems which require special attention are those using a base larger than 10. A good case in point is base 12. Recall that base ten uses 10 one-digit numerals which are 0 through 9, base five uses 5 one-digit numerals from 0 through 4, and base three uses 3 one-digit numerals from 0 through 2. On this basis, it should be clear that base twelve would require 12 one-digit numerals, from 0 through 11. Since there are only 10 Hindu–Arabic numerals, it shall be necessary to invent two additional one-digit numerals to represent ten and eleven. Many authors use a t to represent a ten, and an e to represent an eleven. Therefore, base twelve can be represented by the set of numerals $\{0,1,2,3,4,5,6,7,8,9,t,e\}$. A base 15 numeral system would need 5 additional symbols, and a base 20 system would require 10 additional symbols. The conversion methods which have been used for the previous bases also apply here. The only difficulty will probably occur in becoming familiar with the two new digits. The following examples will aid in gaining that familiarity:

(g) Convert 17 to base twelve.

$$\begin{array}{r} 0 \\ 12\overline{)1} \\ 12\overline{)17} \end{array} \quad \begin{array}{l} -R \quad 1 \\ -R \quad 5 \end{array}$$

—Therefore, $17 = 15_{12}$

(h) Convert 163 to base twelve.

$$\begin{array}{r} 0 \\ 12\overline{)1} \\ 12\overline{)13} \\ 12\overline{)163} \end{array} \quad \begin{array}{l} -R \quad 1 \\ -R \quad 1 \\ -R \quad 7 \end{array}$$

—Therefore, $163 = 117_{12}$

Checking this result,

$$117_{12} = 1 \cdot 12^2 + 1 \cdot 12 + 7$$
$$= 144 + 12 + 7$$
$$= 163$$

(i) Convert 286 to base twelve.

$$\begin{array}{r} 0 \\ 12\overline{)1} \\ 12\overline{)23} \\ 12\overline{)286} \end{array} \quad \begin{array}{l} -R \quad 1 \\ -R \quad 11 \\ -R \quad 10 \end{array}$$

Replacing 11 by e and 10 by t, the desired base twelve numeral is $1et_{12}$ (one e t base twelve). Converting into base ten again,

$$1et_{12} = 1 \cdot 12^2 + 11 \cdot 12 + 10$$
$$= 144 + 132 + 10$$
$$= 286$$

(j) Find the sum of $1t3_{12}$ and 838_{12}.

$$\begin{array}{r} {}^1 1\ t\ 3_{12} \\ 8\ 3\ 8_{12} \\ \hline t\ 1\ e_{12} \end{array}$$

In the above, $t_{12} + 3_{12} = 10 + 3 = 13 = 1 \cdot 12 + 1 = 11_{12}$.

Checking,

$$\begin{array}{l} 1\ t\ 3_{12} = 1 \cdot 12^2 + 10 \cdot 12 + 3 \ \ = \ \ 267 \\ 8\ 3\ 8_{12} = 8 \cdot 12^2 + 3 \cdot 12 + 8 \ \ = 1,196 \end{array} \Big\} 1,463$$

$$t\ 1\ e_{12} = 10 \cdot 12^2 + 1 \cdot 12 + 11 = 1,463$$

(k) Find the product of 314_{12} and $t3_{12}$.

Step one:

$$\begin{array}{r} 3\ {}^1 1\ 4_{12} \\ t\ 3_{12} \\ \hline 9\ 4\ 0 \end{array}$$

Step two:

$$\begin{array}{r} 1_3{}^3 1\ 4_{12} \\ t\ 3_{12} \\ \hline 9\ 4\ 0 \\ 2\ 7\ 1\ 4 \\ \hline 2\ 7,\ t\ 8\ 0_{12} \end{array}$$

In step two above,

$$t_{12} \cdot 4_{12} = 10 \cdot 4 = 40 = 3 \cdot 12 + 4 = 34_{12}$$
$$t_{12} \cdot 3_{12} = 10 \cdot 3 = 30 = 2 \cdot 12 + 6 = 26_{12}$$

Checking,

$$\begin{array}{ll} 314_{12} = 3 \cdot 12^2 + 1 \cdot 12 + 4 = 432 + 12 + 4 & 448 \\ t3_{12} = 10 \cdot 12 + 3 & 123 \\ \hline 27, t80_{12} = 2 \cdot 12^4 + 7 \cdot 12^3 + 10 \cdot 12^2 + 8 \cdot 12 & 1344 \\ & 896 \\ = 41{,}472 + 12{,}096 + 1{,}440 + 96 & 448 \\ = 55{,}104 & \overline{55{,}104} \end{array}$$

Since the latin words for 2 and 10 are *duo* and *decem*, respectively, the base twelve $(2 + 10)$ numeral system is generally referred to as the

duodecimal system. A group has been urging for years to replace the decimal system (base 10) by the duodecimal system. They believe that since 12 contains more factors ($\{1,2,3,4,6,12\}$) than 10 ($\{1,2,5,10\}$), the duodecimal system would be more flexible in working with units of length and fractions. There seems to be little inclination at present to make this change.

In this section, the student has been exposed to the conversion of base ten numerals into other bases. After some practice, which the following exercises shall offer, he should also be able to convert numerals written in different bases to base ten as well as performing arithmetic operations with these numerals. Base two, not included here, shall be discussed separately in the next section.

The commutative and associative properties of addition and multiplication apply for these different base systems. There will be no attempt to generally prove these here; however, the student will be asked to verify specific cases in the exercises.

7.3. Exercises

Convert each of the following numerals, which are written in the bases as indicated, to Hindu–Arabic (base ten) numerals. Check your result occasionally by converting back to the given base.

1. 43_7
2. 26_7
3. 58_9
4. 47_9
5. 536_7
6. 443_7
7. $1,234_6$
8. $4,321_6$
9. 123_4
10. 321_4
11. $5,723_8$
12. $7,046_8$
13. $1,212_3$
14. $1,201_3$
15. $36t_{12}$
16. $t3e_{12}$

17. Write and complete the addition table for base four.
18. Write and complete the multiplication table for base four.
19. Write and complete the multiplication table for base six.
20. Write and complete the addition table for base six.
21. Write and complete the addition table for base nine.
22. Write and complete the multiplication table for base nine.
23. Write and complete the multiplication table for base twelve.
24. Write and complete the addition table for base twelve.

Verify that the associative property is valid for the following numerals:

25. $(67_8 + 35_8) + 45_8 = 67_8 + (35_8 + 45_8)$
26. $42_6 \cdot (35_6 \cdot 13_6) = (42_6 \cdot 35_6) \cdot 13_6$

Perform the operations indicated in the given base for each of the following exercises. Check your result occasionally as demonstrated in examples (j) and (k):

27. $37_8 + 43_8$
28. $148_9 + 364_9$
29. $13_4 \cdot 22_4$
30. $43_7 \cdot 16_7$
31. $121_3 + 212_3$
32. $23_{12} \cdot 4t_{12}$
33. $46_9 \cdot 38_9$
34. $243_6 \cdot 35_6$
35. $48t6_{12} + 3e35_{12}$
36. $213_4 + 322_4 + 131_4$
37. $57_8 \cdot 23_8$
38. $121_3 \cdot 202_3$
39. $4{,}352_6 + 2{,}104_6 + 3{,}410_6 + 5{,}045_6$
40. $253_7 + 4{,}162_7 + 6{,}104_7 + 3{,}523_7 + 5{,}160_7$

7.4. Binary Numeral Systems

The previous sections in this chapter have dealt with numeral systems using many bases other than base ten. The base two numeral system, called the **binary numeral system,** was excluded because it has some important applications which will be discussed later in this section.

As mentioned previously, a positional numeral system uses exactly as many one-digit numerals as the given base. That is, base 10 uses ten one-digit numerals, base five uses five, and base two uses the two one-digit numerals $\{0,1\}$. The different positions of the digits in the binary numeral system are multiples of powers of 2, such as $2^2 = 4$, $2^3 = 8$, $2^4 = 16$, $2^5 = 32$, $2^6 = 64$, etc. The following examples illustrate the conversion of some of these new numerals:

(a) $7 = 4 + 2 + 1$
$= 1 \cdot 2^2 + 1 \cdot 2 + 1$
$= 111_2$

(b) $21 = 16 + 4 + 1$
$= 1 \cdot 2^4 + 0 \cdot 2^3 + 1 \cdot 2^2 + 1$
$= 1{,}011_2$

(c) $10_2 + 11_2 = 101_2$ Since $1_2 + 1_2 = 10_2$

(d) Add $11{,}011_2 + 1{,}101_2$.

$$\begin{array}{r} 1\ 1\overset{1}{0}\overset{1}{1}\ 1\ 1_2 \\ 1\ 1\ 0\ 1_2 \\ \hline 1\ 0\ 1{,}0\ 0\ 0_2 \end{array}$$

This result can be checked by converting to base ten.

$11011_2 = 1 \cdot 2^4 + 1 \cdot 2^3 + 0 \cdot 2^2 + 1 \cdot 2 + 1 = 27$
$1101_2 = 1 \cdot 2^3 + 1 \cdot 2^2 + 0 \cdot 2 + 1\ \ \ \ \ \ \ \ \ \ = 13$
$101{,}000_2 = 1 \cdot 2^5 + 0 \cdot 2^4 + 1 \cdot 2^3 + 0\ \ \ \ \ \ \ \ = 40$

(Sum = 40)

(e) Multiply 11_2 by 101_2.

$$\begin{array}{r} 1\ 0\ 1_2 \\ 1\ 1_2 \\ \hline 1\ 0\ 1 \\ 1\ 0\ 1\ \ \ \\ \hline 1{,}1\ 1\ 1_2 \end{array}$$

Converting to base ten to check the result,

$10 1_2 = 1 \cdot 2^2 + 1\ = 5$
$1 1_2 = 1 \cdot 2 + 1\ = 3$
$1{,}1 1 1_2 = 1 \cdot 2^3 + 1 \cdot 2^2 + 1 \cdot 2 + 1 = 15$

(Product = 15)

(f) Multiply $1{,}011{,}011_2$ by $11{,}011_2$.

$$\begin{array}{r} 1\ 0\ 1\ 1\ 0\ 1\ 1_2 \\ 1\ 1\ 0\ 1\ 1_2 \\ \hline 1\ 0\ 1\ 1\ 0\ 1\ 1 \\ 1\ 0\ 1\ 1\ 0\ 1\ 1\ \ \ \\ 1\ 0\ 1\ 1\ 0\ 1\ 1\ 0\ \ \ \\ 1\ 0\ 1\ 1\ 0\ 1\ 1\ \ \ \ \ \ \ \\ \hline 1\ 0\ 0{,}1\ 1\ 0{,}0\ 1\ 1{,}0\ 0\ 1_2 \end{array}$$

Checking the result by the usual method,

$$1,011,011_2 = 2^6 + 0 + 2^4 + 2^3 + 9 + 2 + 1$$
$$= 64 + 16 + 8 + 2 + 1 \qquad\qquad = 91$$
$$11,011_2 = 2^4 + 2^3 + 0 + 2 + 1$$
$$= 16 + 8 + 2 + 1 \qquad\qquad = 27$$
$$\left.\begin{array}{r}=91\\=27\end{array}\right\}2,457$$
$$100,110,011,001_2 = 2^{11} + 0 + 0 + 2^8 + 2^7 + 0 + 0 + 2^4 + 2^3 + 0 + 0 + 1$$
$$= 2,048 + 256 + 128 + 16 + 8 + 1 \qquad = 2,457$$

(g) Convert 2,457 to binary notation.

```
         0      —R    1
       2)1      —R    0
       2)2      —R    0
       2)4      —R    1
       2)9      —R    1
      2)19      —R    0
      2)38      —R    0
      2)76      —R    1
     2)153      —R    1
     2)307      —R    0
     2)614      —R    0
    2)1228      —R    1
    2)2457
```

Reading the remainders from top to bottom, the equivalent binary numeral is 100,110,011,001. Commas are inserted for ease of reading. This should demonstrate to the student that the repetitive division method can yield the proper result in binary notation, although it is somewhat lengthy at times.

The binary numeral system is the simplest positional system that man can devise since it uses the smallest set of numerals. What can be simpler—any number expressed in binary notation contains some combination of only 0's or 1's. As mentioned previously, a positional system cannot be constructed using a base of 0 or 1. This simple feature of the binary numeral system has several important applications in industry. First, 0 and 1 can be compared to a mechanical switch which is either open (0) or closed (1). This is indicated by the two diagrams of a single pole switch below:

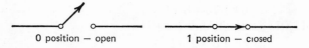

0 position — open   1 position — closed

A simple vacuum tube can be made to behave in the same way as the mechanical switch above. That is, when the tube is not conducting, current does not flow, and this is indicated by a 0. When the tube conducts, current flows, and this is indicated by a 1. A row of tubes can be wired electrically in such a way that each tube can be caused to conduct or not depending upon whether each digit of a given binary numeral is a 0 or a 1. The tube circuits in turn can light small bulbs or not depending upon whether they are conducting. For example, the following binary numerals can be represented by a row of lights as indicated:

(h) $1,011_2$ ○ ● ○ ○   ○—Indicates bulb lighted
(i) $10,100_2$ ○ ● ○ ● ●   ●—Indicates bulb dark

The sum of two one-digit binary numerals are either 0, 1, or 10, depending upon whether they are alike and both zero, unlike, or alike and both 1, in that order. Circuits can be designed to function in accordance with these results. That is, two circuits can be wired electrically in such a way that when both circuits are not conducting (both 0), there is no output ($0 + 0 = 0$). When one is conducting and the other is not (one 0 and the other a 1), then the output can be arranged to be conducting ($0 + 1 = 1 + 0 = 1$). Whenever both circuits are conducting, there is no output (0), but there is an output signal which is "carried" to the circuit of the next higher digit ($1 + 1 = 10_2$).

Consider the addition of $10_2 + 11_2$. The units digits are unlike, so that the circuit can be arranged to indicate a 1, or a conducting circuit. The second digits are alike, so that their circuit can indicate a 0, a nonconducting circuit, and also "carry" a signal to the third digit. In this case, the 2^2 digit is absent, or essentially zero, therefore the signals are unlike and the result is 1. If the 2^2 digit had been a 1, then the signals would have been alike, the 2^2 digit would have changed to a 0, and a 1 signal would have been sent to the 2^3 digit, etc. The diagram below indicates this addition:

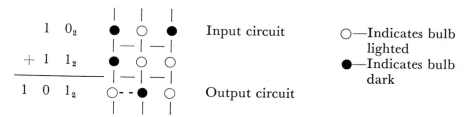

$$\begin{array}{r} 1\ 0_2 \\ +\ 1\ 1_2 \\ \hline 1\ 0\ 1_2 \end{array}$$

Input circuit

Output circuit

○—Indicates bulb lighted
●—Indicates bulb dark

In this way, addition can be performed and either signals emitted or lights energized to signify the result. Without attempting to go into detail, proper circuits can be designed not only to perform the remainder of the arithmetic operations, but most mathematical processes as well.

Although most digital computers operate with binary numerals, it is not necessary to "feed" numbers into the computer in this form. Most computers are designed to receive numerical information in base ten and have circuits which automatically convert this data internally into binary numerals. The binary numeral system does have disadvantages in that many digits are required in order to write large numbers. If it is necessary to find errors or difficulties in the process, the computer operator, or programmer, can have the machine "print out" the results at any step of the process. To make the "print out" more legible, the machine converts the data internally into the **octal** (base 8) or **hexadecimal** (base 16) systems which require fewer digits than the equivalent binary numerals. The final output from the computer is automatically converted to base ten regardless of the base used internally.

The previous explanation was a brief, perhaps excessively so, description of the fundamental principles of the **digital computer.** It is far beyond the scope or interests of this text to further discuss the technicalities of the computer. The subject was introduced for the purpose of demonstrating to the student how numeral systems, and in particular, the binary numeral system, have tremendous applications which may not be at all obvious.

The electronic switches used when the computer was initially developed in 1946 have since been replaced by electronic components, called semiconductors, which are literally thousands of times physically smaller. A modern computer may contain in excess of a million of these components and can perform at unbelievable speeds of 25,000,000 simple additions per second. It is little wonder that a high-speed computer can perform mathematical tasks in seconds which may take many years for a competent mathematician.

The same principle as above is employed in another facet of industry. That is, the repetitive handling of large masses of data. Payrolls, bank statements, vital records, invoices of large corporations, and college registrations are but a few of the tasks performed under the title of **data processing.** The digital computer principle also has applications in such scientific fields as diagnostic medicine, education, game theory, x-ray crystallography, and logic, as well as many others. The computer industry has had phenomenal growth in the last 25 years and shows signs of continuing to do so for many years.

7.4. Exercises

Write each of the following binary numerals as Hindu–Arabic numerals:

1. 110_2
2. $1,010_2$
3. $1,100_2$
4. $10,101_2$
5. $11,011_2$
6. $1,011,011_2$
7. $1,101,101_2$
8. $1,011,011,110_2$

9. Write and complete the addition table for base two.
10. Write and complete the multiplication table for base two.

Perform the indicated operations for each of the following:

11. $11_2 + 11_2$
12. $110_2 + 101_2$
13. $10_2 \cdot 11_2$
14. $101_2 \cdot 10_2$
15. $11,011_2 + 10,101_2$
16. $111,010_2 + 101,101_2$
17. $1,101_2 \cdot 1,011_2$
18. $10,111_2 \cdot 111_2$

Convert each of the following Hindu–Arabic numerals into binary numerals:

19. 13
20. 47
21. 65
22. 97
23. 137
24. 246
25. 483
26. 956

27. Complete the following sentence: All even numbers which are written in binary notation can be easily identified because _____.

28. A binary numeral is easily converted into octal (base 8) notation by writing each triplet in base 10. Thus, $101{,}110_2 = 56_8$. Verify this result by converting both sides of the equality into base 10.

7.5. Summary

This chapter has been devoted to a study of positional numeral systems having bases other than ten. The Babylonian system was introduced to provide historical evidence of a positional system using the unusual base of 60. Additional motivation was supplied to show the feasibility of counting in terms other than the traditional ten and powers of ten.

The commutative and associative properties, defined for base ten numerals, were also found to apply to numeral systems using other bases. Although not discussed in this chapter, the identity elements for addition and multiplication are 0 and 1, respectively, and the distributive property is valid for bases other than ten.

Numerous exercises were provided for the student to convert from numbers written in numerals in various base systems as well as the conversion from base ten to various base notations.

Finally, an attempt was made to describe how binary numerals are used in the computer. Hopefully, the student is able to gain some insight into its workings from these few paragraphs. Unfortunately, the digital computer is not based upon a single simple concept and therefore cannot be simply described. Those students having some knowledge in electricity should have no trouble finding much reading matter at any level on this subject.

7.5. Review Exercises

Write each of the following Hindu–Arabic numerals as a Babylonian numeral:

1. 56
2. 671

Write each of the following Babylonian numerals as Hindu–Arabic numerals:

3. <<<|||||

4. <<|$\genfrac{}{}{0pt}{}{<<}{<<}$||

Write each of the following sums as numerals in the same base as the indicated powers:

5. $3 \cdot 4^2 + 2 \cdot 4 + 1$
6. $8 \cdot 9^4 + 0 \cdot 9^3 + 2 \cdot 9 + 5$
7. $4 \cdot 7^3 + 4$
8. $2 \cdot 12^3 + 11 \cdot 12^2 + 7 \cdot 12 + 10$
9. $1 \cdot 2^5 + 1 \cdot 2^3 + 1$
10. $3 \cdot 5^5 + 2 \cdot 5^4$
11. $10 \cdot 12^4 + 3$
12. $5 \cdot 6^5 + 3 \cdot 6^2$

Convert each of the following numerals, written in the bases as indicated, into Hindu–Arabic numerals:

13. 42_5
14. 26_7
15. 312_4
16. 403_6
17. $1,011_2$
18. $4t_{12}$
19. 88_9
20. 263_8
21. $1,020_3$
22. $4,100_5$
23. $te0_{12}$
24. $11,010_2$
25. $3,712_8$
26. $24,030_5$
27. $60,003_7$
28. $100,001,010_2$

Convert each of the following Hindu–Arabic numerals into equivalent numbers written in the base as indicated:

29. 43, base 4
30. 82, base 9
31. 143, base 12
32. 77, base 2
33. 223, base 5
34. 223, base 6
35. 223, base 7
36. 223, base 2
37. 223, base 8
  (Hint: see Exercise 28, Section 7.4)
38. 2,431, base 12
39. 700, base 3
40. 500, base 15

Perform the indicated operation on each of the following:

41. $24_7 + 35_7$
42. $23_7 \cdot 32_7$
43. $53_8 + 146_8$
44. $101_2 \cdot 1,110_2$
45. $121_3 + 2,121_3$
46. $26_{12} + 35_{12}$
47. $32_4 \cdot 23_4$
48. $2,314_5 + 132_5$
49. $63_9 \cdot 24_9$
50. $2,303_4 + 212_4$
51. $1t_{12} \cdot 24_{12}$
52. $110,011,101_2 + 101,101,011_2$

Convert each of the given numerals into an equivalent number written in the indicated base:

53. $25_8 = (\quad)_3$
54. $36_{12} = (\quad)_2$
55. $1{,}243_5 = (\quad)_4$
56. $666_7 = (\quad)_6$

Verify that the distributive property holds by calculating each side of the following statements:

57. $23_4(12_4 + 31_4) = 23_4 \cdot 12_4 + 23_4 \cdot 31_4$
58. $52_7(34_7 + 15_7) = 52_7 \cdot 34_7 + 52_7 \cdot 15_7$
59. Write and complete the multiplication table for base three.
60. A binary numeral is easily converted into hexadecimal (base 16) notation by writing every four digits in base 10. Thus, $1{,}011{,}001_2 = 59_{16}$. Verify this result by converting both sides of the equality into base 10.

Table of Squares and Square Roots

No.	Square	Square root	No.	Square	Square root
1	1	1.000	51	2,601	7.141
2	4	1.414	52	2,704	7.211
3	9	1.732	53	2,809	7.280
4	16	2.000	54	2,916	7.348
5	25	2.236	55	3,025	7.416
6	36	2.449	56	3,136	7.483
7	49	2.646	57	3,249	7.550
8	64	2.828	58	3,364	7.616
9	81	3.000	59	3,481	7.681
10	100	3.162	60	3,600	7.746
11	121	3.317	61	3,721	7.810
12	144	3.464	62	3,844	7.874
13	169	3.606	63	3,969	7.937
14	196	3.742	64	4,096	8.000
15	225	3.873	65	4,225	8.062
16	256	4.000	66	4,356	8.124
17	289	4.123	67	4,489	8.185
18	324	4.243	68	4,624	8.246
19	361	4.359	69	4,761	8.307
20	400	4.472	70	4,900	8.367
21	441	4.583	71	5,041	8.426
22	484	4.690	72	5,184	8.485
23	529	4.796	73	5,329	8.544
24	576	4.899	74	5,476	8.602
25	625	5.000	75	5,625	8.660
26	676	5.099	76	5,776	8.718
27	729	5.196	77	5,929	8.775
28	784	5.292	78	6,084	8.832
29	841	5.385	79	6,241	8.888
30	900	5.477	80	6,400	8.944
31	961	5.568	81	6,561	9.000
32	1,024	5.657	82	6,724	9.055
33	1,089	5.745	83	6,889	9.110
34	1,156	5.831	84	7,056	9.165
35	1,225	5.916	85	7,225	9.220
36	1,296	6.000	86	7,396	9.274
37	1,369	6.083	87	7,569	9.327
38	1,444	6.164	88	7,744	9.381
39	1,521	6.245	89	7,921	9.434
40	1,600	6.325	90	8,100	9.487
41	1,681	6.403	91	8,281	9.539
42	1,764	6.481	92	8,464	9.592
43	1,849	6.557	93	8,649	9.644
44	1,936	6.633	94	8,836	9.695
45	2,025	6.708	95	9,025	9.747
46	2,116	6.782	96	9,216	9.798
47	2,209	6.856	97	9,409	9.849
48	2,304	6.928	98	9,604	9.899
49	2,401	7.000	99	9,801	9.950
50	2,500	7.071	100	10,000	10.000

Glossary

Absolute value: The distance that a number is from zero without regard to direction. The symbol is vertical parallel bars, | |. Example: $|-7| = 7$

Additive inverse: If the sum of two numbers is zero, the identity element, these numbers are said to be the additive inverses of the other. Example: Since $5 + (-5) = 0$, 5 and -5 are additive inverses of each other.

Algorithm: A mathematical process which has been shortened and simplified.

Associative: An operation is associative if it can be performed on two different groupings of elements, in the same order, with the same results. Example:
$$(7 + 3) + 5 = 7 + (3 + 5)$$
$$(8 \cdot 4) \cdot 9 = 8 \cdot (4 \cdot 9)$$
$$A \cup (B \cup C) = (A \cup B) \cup C$$

Average: The average of a set of n quantities is defined as the sum of these n quantities divided by n. Example: The average of 7, 4, 5, 9, and 15 is $\dfrac{7 + 4 + 5 + 9 + 15}{5} = 8.$

Base (of exponents): The number which has been raised to an exponent.
  (of numeral system): The number around which a positional numeral system is built.
Binary: An adjective meaning two.

Cardinality: The number of elements in a set.
Closed, closure: A set is closed when any two numbers in a given set are combined by a given operation, and the result is always another member of the given set.
Commutative: An operation upon two elements of a set is commutative if the operation can be performed with the elements in different order to get the same result.
Constant: Symbols having only one value.

Dense: A number system is dense if given any two numbers in the system, there is at least one number between them on the number line.
Difference: The result of the subtraction of two numbers.
Directed distance: The distance between numbers in a specified direction.
Distributive: A property which is a connecting link between addition and multiplication, as well as subtraction and multiplication.
Dividend: The number which is divided **into** in a division operation.
Division: The number which one divides **by** in a division operation.
Domain: The replacement set of a variable.
Double negative: The negative of the negative of a number.
Duodecimal: A numeral system based upon 12.

Equalities: Statements in which the quantities compared are equal.
Equation: An equality which is an open sentence.
Exactly divisible: When two quantities are divided with a remainder of zero, one quantity is exactly divisible by the second.
Expanded notation: Writing numbers as the sum of the values of the digits. That is, not in positional notation.
Exponent: A number indicating the number of factors of the given base.

Factor (verb): The process of finding the factors of a given number.
(noun): When numbers are multiplied together to produce a product, each number is a factor of the product.

Field: A number system satisfying commutative, associative, identity elements, inverse elements, and the closure property for two operations, and a distributive property connecting the two operations.

Fraction: Numbers of the form $\frac{a}{b}$, $b \neq 0$, where a and b are integers, are called simple fractions. When a and/or b are rational numbers, it is called a compound fraction.

Graph: The pictorial representation of any numerical data.

Greatest common factor (GCF): The largest factor which two or more numbers have in common.

Hexadecimal: A numeral system based upon 16.

Identity: An equation which is always true for all replacements of the variable.

Identity element: An element which when combined with each of the elements of a given set, for a given operation, does not change the identity of each of the elements. For example:

$7 + 0 = 7,$ 0 is the identity element for addition
$6 \cdot 1 = 6,$ 1 is the identity element for multiplication
$A \cup \emptyset = A,$ $\emptyset$ is the identity element for union

Indeterminate: Any expression whose value cannot be uniquely determined, specifically $\frac{0}{0}$.

Index: The indicated root of a radical. (Plural: indices.)

Inequality: A statement in which the quantities compared are unequal.

Integers: $J = \{\ldots, -3, -2, -1, 0, 1, 2, 3, \ldots\}$

Inverse: When two elements, combined with respect to a given operation, produce the identity element for that operation, the two elements are inverse elements of each other for that given operation. Example:

$7 + (-7) = 0$ 7 and -7 are additive inverse elements
$7 \cdot \frac{1}{7} = 1$ 7 and $\frac{1}{7}$ are multiplicative inverse elements

Irrational: Any number which cannot be expressed as the quotient of two integers.

Least common denominator (LCD): The least common multiple of two or more denominators.
Least common multiple (LCM): The smallest natural number which is a multiple of all elements of any finite set of natural numbers.

Minuend: The given number in a subtraction operation.
Multiple: The multiple of a number is that number multiplied by any integer.

Natural Numbers: $N = \{1,2,3,\ldots\}$
Negative: The negative integers are $\{\ldots,-3,-2,-1\}$. The negative of a number transfers a number to the opposite side of zero. Thus, $-(+3) = -3$ and $-(-7) = +7$.
Number line: A line on which a set of numbers can be graphically represented.
Numeral: Symbols which represent numbers.
Numerator: In any fraction $\frac{a}{b}$, the numerator is a.

Octal: A numeral system based upon 8.
Open sentence: A statement which is neither true nor false, but depending upon the choice of the replacements of the variable.
Operation: Generally, a mathematical process. When two numbers are combined in some way to produce a third, not necessarily different, the process is binary. When a process is performed on a single number to produce a second, not necessarily different, the process is unary.
Opposite: A synonym for negative.
Ordinal: A number which represents the relative position of objects in a given order. Example: second, third, etc.

Polynomial: A particular kind of algebraic expression.
Positional: A numeral system in which each digit has place value and face value.

Power: A number raised to a given power has that many factors of the base.
Prime number: A natural number greater than 1 having no factors other than 1 and itself.
Prime, relatively: See relatively prime.
Product: The result of the multiplication of two or more numbers.

Quotient: The result of the division of two numbers.

Radical: An expression consisting of a radical sign ($\sqrt{}$), the radicand, and possibly an index.
Radicand: The expression inside a radical sign.
Rational (number): Any number which can be expressed as the quotient of two integers.
Real (number): Any number which can be rational or irrational.
Reciprocal: If b is a nonzero number, then $\frac{1}{b}$ is its reciprocal.
Relation: A comparison of two quantities such as $=$, $\neq$, $>$, or $<$.
Relatively prime: Two numbers having no factors in common, other than 1.
Remainder: The amount left over when two numbers do not divide evenly.
Replacement set: The set of replacements for a given variable.
Root: The nth root of a given number is one of the n equal factors of that number.

Satisfy: When the replacement value of a variable results in a true sentence, it "satisfies" the sentence.
Scientific notation: Writing numbers as a number between one and ten times an appropriate integral power of ten.
Set builder notation: A detailed method of specifying a solution set.
Sets: A group of objects having some recognizable property in common.
Simple grouping: A numeral system in which each symbol has one value regardless of its position in the numeral.
Subset: If each element of any set A is also an element of a set B, then set A is called a subset of set B.

Subtrahend: The number to be subtracted in a subtraction operation.

Sum: The result of an addition operation.

Unary: An adjective meaning one.

Union: The union of any two sets A and B is the set which consists of all elements which are either in A or B, or both A and B.

Variable: Any symbol which can be replaced by more than one numeral.

Whole number: $W = \{0,1,2,3,\ldots\}$

Answers for Selected Odd-Numbered Exercises

Chapter 1

Section 1.1

1. {Spring,Summer,Fall,Winter}
3. {penny,nickel,dime,quarter,half-dollar,dollar}
5. {d,e,f,g,h,i,j}
7. {Cities in the U.S. with a population exceeding 2 million people} (Based on 1970 census)
9. {Days of the week}
11. {States in the U.S. bordering the Pacific Ocean}
13. 365, except leap year
15. 50
17. 5
19. ϱϱ∩∩∩||||
21. 𓋹ϱϱϱϱϱϱϱϱϱ∩∩∩∩∩∩

276 Pre-Algebra

23. 35
25. 200,023
27. CCLXIII
29. XCIV
31. 62
33. 590

Section 1.2

1. 204
3. 50,505
5. 4,063,029
7.
9. CMXLIV
11. 0; 10,000
13. 8; 10,000
15. 3
17. 0
19. 4
21. infinite
23. finite
25. infinite
27. {5,6,7,...,175}
29. {2,7,12,17}
31. {0,1,2,...,16}

Section 1.3

1. {2,3,5,8,11}
3. {3,6,9,...,84}
5. {4,5,6,...,13}
7. {a,e,r,s,w}
9. 4, 7
11. 3
13. 100
15. 7
23. Binary

Section 1.4

1.
3.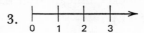
5.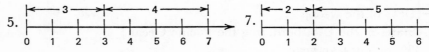
7.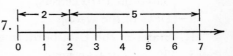
9. $600 + 40 + 7$
11. $20,000 + 800 + 7$
13. $(20 + 4) + (300 + 60 + 5)$
    $= 300 + (20 + 60) + 4 + 5$
    $= 300 + 80 + 9$
    $= 389$
15. 2,853
17. 37
19. 33
21. 30,065
23. 51,580

Answers 277

Section 1.5

1. 8
3. None exists
5. 21
7. 3,384
9. 79
11. 24,586
13. 6
15. 11
17. None exists
19. 2
21. {0}, {0}
23. {2,4,6,8,...}, no

Section 1.6

1. 189
3. 12,936
5. 162,936
7. 0
9. 0
11. 61
13. 46
15. 5
17. 41
19. 16
21. 337
23. 13
25. 9
27. 16
29. 0

Section 1.7

1. >
3. >
5. =
7. >
9. >
11. False
13. False
15. $8 < n < 13$
17. $0 < n < 9$
19. $17 \leq n \leq 83$
21. Less than or equal to

23.

Section 1.8

1. 2,131
3. 2,069
5. 200,202
7. {9,10,11,...,92}
9. {a,c,d,g,r,t,z}
11. {17,18,19,...,83}
13. $400 + 30 + 7$
15. 1,702
17. 0
19. 71
21. 7
23. $5 < n < 73$ or $6 \leq n \leq 72$
25. $n > 43$
27. Inequality
29. Cardinality
31. Binary operation
33. Hindu–Arabic numerals
35. Positional numeral system; simple grouping

Chapter 2

Section 2.1

1. 12	3. 0	5. 15	7. 11
9. −8	11. =	13. =	15. <
17. >	19. >	21. >	23. <
25. False	27. True	29. False	31. True

Section 2.2

1. 2	3. 4	5. −2	7. 4
9. 8	11. 10	13. −1	15. −14
17. −1	19. −2	21. 0	23. 9
25. −4	27. −5	29. −2	31. −12
33. −7	35. 2	37. −11	39. −6
41. 6	43. 3	45. −5	47. −3

Section 2.3

1. −3	3. 5	5. −3	7. 8
9. 19	11. 15	13. −18	15. −3
17. −3	19. 4	21. −2	23. −19
25. 10	27. −12	29. 4	31. −7
33. −403	35. −607		

Section 2.4

1. −5	3. −13	5. 11	7. 4
9. 3	11. −26	13. 9	15. −5
17. −4	19. 3	21. −5	23. 9
25. −4	27. −6	29. −20	31. 6
33. 7	35. 1	37. 5	39. 14
41. −7	43. −7	45. −14	47. 198

49. −188 51. −5,449 53. −11 55. =
57. >   59. <   61. =   63. =
65. <   67. =   69. <   71. =

Section 2.5

1. −28   3. −24   5. 18   7. −36
9. 30   11. 72   13. −60 15. −90
17. 21   19. 35   21. 6   23. −25
25. −7   27. −21 29. 1   31. −68
33. −10 35. 25   37. −5   39. 0
41. <   43. =   45. =   47. =

Section 2.6

1. −12   3. −14   5. −72   7. 9
9. 33   11. 32   13. −32 15. −26
17. −48 19. 0   21. −19 23. −76
25. 430 27. 464 29. −1,160 31. 60
33. 144 35. −136
37. 1 · 18, 2 · 9, 3 · 6
39. 1 · 42, 2 · 21, 3 · 14, 6 · 7
41. 3   43. 1
45. 5   47. 2   49. 13

Section 2.7

1. −3   3. 8   5. −3   7. 2
9. 1   11. −17 13. −3   15. 18
17. 27   19. −33 21. 0   23. −50
25. >   27. =   29. <   31. >
33. =   35. <
37. 1 · 24, 2 · 12, 3 · 8, 4 · 6
39. 1 · 54, 2 · 27, 3 · 18, 6 · 9
41. 4   43. 6
45. 3   47. Commutative   49. ≥
51. Sometimes

Chapter 3

Section 3.1

1. 8
3. $\dfrac{4}{7}$
5. $\dfrac{14}{5}$
7. 1
9. $\dfrac{6}{5}$
11. 2
13. -3
15. $\dfrac{8}{21}$
17. $\dfrac{40}{63}$
19. $\dfrac{-8}{27}$
21. $\dfrac{15}{56}$
23. $\dfrac{32}{63}$
25. $-\dfrac{32}{63}$
27. $\dfrac{10}{27}$
29. -12
31. $\dfrac{225}{49}$
33. $\dfrac{1}{4}$
35. $\dfrac{1}{-3}$
37. $\dfrac{-1}{8}$
39. 2

Section 3.2

1. $\dfrac{7}{4}$
3. -3
5. $\dfrac{7}{8}$
7. $\dfrac{2}{3}$
9. $\dfrac{5}{9}$
11. $\dfrac{-3}{2}$
13. $\dfrac{-7}{2}$
15. $\dfrac{27}{49}$
17. $\dfrac{-3}{8}$
19. $\dfrac{-3}{7}$
21. $\dfrac{3}{7}$
23. $\dfrac{-2}{7}$
25. $\dfrac{24}{25}$
27. 12
29. -9
31. -45
33. -35
35. 21
37. 9
39. -5
41. 3
43. 4
45. 3
47. -17
49. 5

Section 3.3

1. $\dfrac{3}{7}$
3. $\dfrac{6}{35}$
5. $\dfrac{-5}{6}$
7. $\dfrac{-21}{2}$

9. $\dfrac{63}{4}$ 11. $\dfrac{-86}{7}$ 13. 1 15. 99

17. $\dfrac{3}{10}$ 19. $\dfrac{3}{5}$ 21. $\dfrac{21}{32}$ 23. $\dfrac{81}{55}$

25. 4 27. $\dfrac{108}{25}$ 29. $\dfrac{4}{21}$ 31. $\dfrac{189}{23}$

Section 3.4

1. $\dfrac{9}{2}$ 3. $\dfrac{2}{45}$ 5. $\dfrac{-1}{14}$ 7. $\dfrac{21}{32}$

9. $\dfrac{17}{25}$ 11. $\dfrac{-20}{3}$ 13. $\dfrac{4}{3}$ 15. $\dfrac{11}{15}$

17. $\dfrac{7}{12}$ 19. -1 21. -4 23. 9

25. $\dfrac{21}{2}$ 27. 4 29. -38 31. $\dfrac{27}{35}$

33. -4 35. $\dfrac{1}{2}$ 37. $\dfrac{1}{42}$ 39. -15

Section 3.5

1. -4 3. -3 5. 4 7. $\dfrac{14}{9}$

9. $\dfrac{-5}{14}$ 11. $\dfrac{15}{2}$ 13. $\dfrac{1}{39}$ 15. -63

17. $\dfrac{7}{27}$ 19. $\dfrac{-26}{35}$ 21. $\dfrac{4}{7}$ 23. $\dfrac{15}{22}$

25. 72 27. 176 29. $39\dfrac{139}{314}$

Section 3.6

1. $2 \cdot 2 \cdot 2 \cdot 3$ 3. $2 \cdot 7 \cdot 7$ 5. $2 \cdot 3 \cdot 3 \cdot 5 \cdot 7$

7. $7 \cdot 11 \cdot 13$ 9. $3 \cdot 3 \cdot 3 \cdot 5 \cdot 7$

282   Pre-Algebra

11. $3 \cdot 3 \cdot 5 \cdot 13 \cdot 13$
13. 63
15. 18
17. 105
19. 660
21. 1,350
23. 1,080
25. 24
27. 36
29. 1,950
31. 975
33. 252
35. 280
37. 2, 3, 5, 7, 11, 13, 17, 19, 23, 29, 31, 37, 41, 43, 47

Section 3.7

1. 1
3. $\dfrac{3}{5}$
5. $\dfrac{3}{4}$
7. $\dfrac{-2}{3}$

9. $\dfrac{5}{8}$
11. $\dfrac{2}{3}$
13. $\dfrac{29}{28}$
15. $\dfrac{118}{105}$

17. $\dfrac{7}{48}$
19. $\dfrac{-1}{60}$
21. $\dfrac{31}{126}$
23. $\dfrac{-55}{247}$

25. $\dfrac{-67}{270}$
27. $\dfrac{-79}{352}$
29. $\dfrac{51}{8}$
31. $\dfrac{45}{7}$

33. $\dfrac{-19}{6}$
35. $\dfrac{-17}{5}$
37. $\dfrac{64}{105}$
39. $\dfrac{77}{90}$

41. $\dfrac{-47}{126}$
43. $\dfrac{216}{7}$
45. $\dfrac{-40}{21}$
47. $\dfrac{81}{56}$

Section 3.8

1. $>$
3. $>$
5. $=$
7. $>$
9. $<$
11. $=$
13. $>$
15. $<$
17. $>$
19. $<$
21. $\dfrac{13}{16}$
23. $\dfrac{-26}{45}$

25. $\dfrac{-15}{28}$
27. $\dfrac{113}{266}$
29. $\dfrac{9}{16}, \dfrac{10}{16}, \dfrac{11}{16}$

31. $\dfrac{57}{96}, \dfrac{58}{96}, \dfrac{59}{96}$
33. $\dfrac{-1}{6}, \dfrac{1}{3}, \dfrac{8}{21}, \dfrac{1}{2}, \dfrac{4}{7}$
35. $\dfrac{3}{5}, \dfrac{7}{11}, \dfrac{2}{3}, \dfrac{5}{7}$

Section 3.9

1. $\dfrac{3}{28}$　　　3. $\dfrac{2}{15}$　　　5. $\dfrac{12}{7}$　　　7. $\dfrac{-10}{3}$

9. $\dfrac{5}{27}$　　　11. $\dfrac{-1}{15}$　　　13. $\dfrac{-3}{5}$　　　15. $\dfrac{-7}{15}$

17. $\dfrac{25}{24}$　　　19. $\dfrac{1}{4}$　　　21. $\dfrac{4}{3}$　　　23. $\dfrac{-5}{24}$

Section 3.10

1. $\dfrac{3}{4}$　　　3. $\dfrac{-4}{3}$　　　5. $\dfrac{-20}{27}$　　　7. $\dfrac{35}{13}$

9. 583　　　11. $\dfrac{2}{63}$　　　13. $\dfrac{-24}{5}$　　　15. $\dfrac{-20}{17}$

17. $\dfrac{3}{28}$　　　19. -1　　　21. $\dfrac{62}{105}$　　　23. $\dfrac{23}{56}$

25. $\dfrac{1}{210}$　　　27. 42　　　29. $\dfrac{7}{3}$　　　31. $\dfrac{-23}{420}$

33. $\dfrac{3}{2}$　　　35. $\dfrac{4}{7}$　　　37. $2 \cdot 3 \cdot 7 \cdot 7$　　　39. $3 \cdot 3 \cdot 5 \cdot 5 \cdot 7$

41. -24　　　43. 39　　　45. 120　　　47. 1,260

49. 1,176　　　51. 1,386　　　53. $\dfrac{4}{15}$　　　55. $\dfrac{12}{5}$

57. $\dfrac{-9}{5}$　　　59. $<$　　　61. $<$

Chapter 4

Section 4.1

1. 2^3　　　3. $7^3 4^4$　　　5. $\left(\dfrac{-4}{9}\right)^4$　　　7. $(-3)^3 \cdot 6^5$

284   Pre-Algebra

9. 3^{11} 11. 5^9 13. $\left(\dfrac{-2}{3}\right)^8$ 15. 5^4

17. -8^2 19. -5^9 21. 1 23. $\dfrac{5^{11}}{6^5}$

25. $8^3 \cdot 7^2$ 27. $\dfrac{-1}{13^4}$ 29. 144 31. 256

33. $\dfrac{9}{64}$ 35. -243 37. -324 39. $\dfrac{-125}{216}$

41. $8 \cdot 10^2 + 3 \cdot 10 + 4$
43. $4 \cdot 10^4 + 8 \cdot 10^3 + 6 \cdot 10^2 + 8 \cdot 10 + 9$

Section 4.2

1. 5^6 3. 8^{12} 5. -7^{15} 7. $5^3 \cdot 7^3$

9. $3^{12} \cdot 8^6$ 11. 3^{14} 13. $2^{18} \cdot 3^{24} \cdot 4^{30}$ 15. $\dfrac{3^4}{4^8}$

17. $\dfrac{(-6)^6}{7^{24}}$ 19. $\dfrac{1}{8^{15}}$ 21. $2^3 \cdot 3^{30} \cdot 4^{45}$ 23. $\dfrac{-1}{3^{13}}$

Section 4.3

1. ± 9 3. ± 11 5. ± 1 7. ± 15
9. 5 11. -2 13. -6 15. -6
17. 2 19. 4 21. 2 23. 2
25. 6 27. 6 29. 7 31. 17
33. $3\sqrt{2}$ 35. $6\sqrt{2}$ 37. $-3\sqrt[3]{3}$ 39. $4\sqrt{3}$
41. $4\sqrt{2}$ 43. $5\sqrt{3}$ 45. $3\sqrt{6}$ 47. $2\sqrt[3]{2}$

Section 4.4

1. $4\sqrt{3}$ 3. $-12\sqrt{6}$ 5. $-3\sqrt{3} - \sqrt{2}$ 7. $8\sqrt[3]{6} - 4\sqrt{6}$

9. $6\sqrt{2}$ 11. $-42\sqrt{2}$ 13. $\dfrac{\sqrt{3}}{2}$ 15. $\dfrac{-2\sqrt{2}}{3}$

17. $\dfrac{\sqrt{3}}{2}$ 19. $6\sqrt{6}$ 21. 7 23. 18
25. $2\sqrt{6}$ 27. 12 29. $4\sqrt{2}$ 31. $2\sqrt{3}-3\sqrt{2}$
33. $6-4\sqrt{3}$ 35. $-36+12\sqrt{15}$ 37. $6\sqrt{15}+60$
39. $-24+48\sqrt{6}$

Section 4.5

1. $2\sqrt{2}$ 3. $\dfrac{3\sqrt{2}}{4}$ 5. $\dfrac{\sqrt{2}}{4}$ 7. $2\sqrt[3]{9}$

9. $\dfrac{\sqrt{3}}{3}$ 11. $\dfrac{\sqrt[3]{6}}{2}$ 13. $\dfrac{-2\sqrt[3]{25}}{5}$ 15. $\dfrac{\sqrt[3]{18}}{3}$

17. $\dfrac{16}{3}$ 19. 75 21. $\dfrac{-3\sqrt{6}}{32}$ 23. $\dfrac{\sqrt{5}}{25}$

25. $\dfrac{5\sqrt{5}}{12}$ 27. $\dfrac{7\sqrt{3}}{30}$ 29. $\dfrac{-\sqrt{6}}{3}$ 31. $\dfrac{-19\sqrt{3}}{4}$

Section 4.6

1. 10 3. $3\sqrt{10}$ 5. $6\sqrt{6}$ 7. 33

9. 3 11. 14 13. $\dfrac{21}{64}$ 15. $\dfrac{119}{13}$

17. $\dfrac{59}{6}$ 19. $3+6\sqrt{3}$

Section 4.7

1. 4^3 3. -5^4 5. 7^8 7. 8^{12}

9. $\dfrac{1}{5^4}$ 11. $(-4)^3$ 13. $3^{12}\cdot 4^{12}\cdot 5^4$ 15. $3^3(-4)^2\cdot 5^4$

17. 6 19. -4 21. $2\sqrt{5}$ 23. $8\sqrt[3]{5}$

25. $\sqrt{3}$
27. $2\sqrt{3}$
29. $\dfrac{\sqrt{6}}{3}$
31. $\dfrac{\sqrt{3}}{3}$

33. $9\sqrt{3}$
35. $\sqrt[3]{2}$
37. $\dfrac{-\sqrt{5}}{6}$
39. $\dfrac{7\sqrt{6}}{12}$

41. $\dfrac{13\sqrt{2}}{6}$
43. $28 - 42\sqrt{2}$
45. 11
47. $3\sqrt{5}$

49. $24\sqrt{3}$
51. 54
53. $-40\sqrt{5}$
55. $-150\sqrt{6}$

57. $3\sqrt{2}$
59. 7
61. -6
63. $5\sqrt{5}$

Chapter 5

Section 5.1

1. $\dfrac{1}{4^3}$
3. $\dfrac{1}{7}$
5. $\left(\dfrac{5}{3}\right)^3$
7. -8^4

9. 4
11. $\dfrac{1}{3}$
13. $\dfrac{1}{10^4}$
15. 10^2

17. 10^{-2}
19. $3^6 \cdot 5^3$
21. 10^6
23. 10^3

25. 2^{-12}
27. $7^{-3} \cdot 4^{-6}$
29. $8^{12} \cdot 5^{-9}$

Section 5.2

1. $4 \cdot 10^{-1} + 9 \cdot 10^{-2} = 0.49$
3. $6 \cdot 10^{-1} + 7 \cdot 10^{-3} = 0.607$
5. $3 \cdot 10^{-4} = 0.0003$
7. $1 \cdot 10^{-4} + 4 \cdot 10^{-5} = 0.00014$

9. $\dfrac{7}{10^2}$
11. $\dfrac{437}{10^4}$
13. $\dfrac{6}{10^4}$
15. $\dfrac{732}{10^5}$

17. 0.125
19. 0.06
21. 0.085
23. 0.65

25. $0.111\overline{1}$
27. $0.27\overline{27}$
29. $0.291\overline{6}$
31. $0.\overline{238095}$

Section 5.3

1. $57 \cdot 10^{-2}$
3. $91 \cdot 10^{-3}$
5. $2{,}081 \cdot 10^{-4}$

7. $85{,}003 \cdot 10^{-3}$
9. 713
11. 31.6

13. 6.83000
19. 400
25. 0.00324
31. 0.2411

15. 0.9073000
21. 9.2415
27. 4,000,000

17. 4
23. 6.32
29. -10.2714

Section 5.4

1. $3 \cdot 10^{-2}$
7. 7.6×10^4
13. 5.008×10^{-2}
19. 8,050
25. 0.0004
31. 0.248

3. $8 \cdot 10^2$
9. $4 \cdot 10^{-5}$
15. 6.8×10^{16}
21. 0.0000632
27. 40

5. 7.31×10^{-3}
11. 9.6×10^7
17. 0.0403
23. 7,835,200,000,000
29. 1,920

Section 5.5

1. 2.449
9. 4.4721

3. 3.317
11. 5.6568

5. 3.742
13. 9.8995

7. 5.916
15. 5.1962

Section 5.6

1. N
9. W

3. Q
11. N

5. Q
13. J

7. N
15. Q

Section 5.7

1. $\dfrac{1}{7^4}$

3. $\dfrac{1}{-4^3}$

5. -8^3

7. $\dfrac{1}{9^3}$

9. 3^{-3}

11. 4^{-8}

13. 7^6

15. $4^{-7} \cdot 6^6$

17. 0.0377

19. 0.809

21. $\dfrac{837}{10^4}$

23. $\dfrac{37}{10^5}$

25. 0.35

27. 0.4375

29. $0.\overline{45}$

31. 0.1225
33. $7 \cdot 10^{-2}$
35. $3,405 \times 10^{-4}$
37. 73.4
39. 497,800
41. 42
43. 2.46
45. 0.000576
47. 0.05453
49. 8.1×10^{-3}
51. 7.8×10^{-1}
53. 6.007×10^{-5}
55. 7.035×10^{-4}
57. 0.0074
59. 0.000,003,003
61. 1,200,000
63. 3,000,000,000
65. 3.873
67. 2.846
69. 11.1804
71. 10.3923

Chapter 6

Section 6.1

1. 6^2
3. $9 - 2$
5. $\dfrac{7}{5}$
7. $2 - 8$
9. $2 \cdot 7 + 8$
11. $2 \cdot 9 - 3 \cdot 8$
13. $-(4 + 7)$
15. $\dfrac{7}{3} + 2$
17. $(-8)^2$
19. $3^3 + 5^3$
21. $(2^3)^2 + (5^3)^2$
23. $\sqrt{4} + \sqrt{7}$
25. $2x$
27. xy
29. $x - 8$
31. $2x - 3$
33. $3x^2 - 9$
35. $4x + (-3)$ or $4x - 3$
37. $(x - y)^2$
39. $3x - 4y$
41. $3(x + y)^2$
43. xy^2
45. $3x^2y^3$
47. $2x^3 + 6x - 11$

Section 6.2

1. $3a$
3. $3x$
5. (a) $x + 2$
   (b) $x - 5$
7. $8p$
9. $x - 1$
11. $x + 2$
13. $x - 2$
15. $x + 2, x + 4$
    or $x - 2, x - 4$
17. rt
19. (a) $x + 6$
    (b) $x + (x + 6)$
21. $20x$
23. $4x$

25. $5n$ 27. $25q$ 29. $\dfrac{n}{2}$

31. $\dfrac{5q}{2}$ 33. $\dfrac{m}{60}$ 35. $\dfrac{n}{20}$

37. $\dfrac{q}{4}$ 39. (a) $2w + 6$ 41. (a) $2x - 26$

(b) $\dfrac{l - 6}{2}$ (b) $x - 4$, $2x - 30$

(c) $\dfrac{y + 26}{2}$

43. (a) $4s$ 45. (a) $2l + 2\left(\dfrac{l + 3}{2}\right)$ 47. x, $x + 4$, $2x$

(b) s^2 (b) $l \cdot \left(\dfrac{l + 3}{2}\right)$

Section 6.3

1. True 3. False 5. True
7. True 9. Identity
11. Conditional equation 13. Identity
15. Conditional equation 17. $\{6\}$

19. $\{-3\}$ 21. $\{-18\}$ 23. $\left\{\dfrac{2}{3}\right\}$

25. $\left\{\dfrac{3}{2}\right\}$ 27. $\emptyset$ 29. R (all real numbers)

31. $\emptyset$ 33. $\left\{\dfrac{8}{3}\right\}$ 35. $\emptyset$

37. $\left\{\dfrac{6}{7}\right\}$ 39. $\left\{-8, -3, 0, \dfrac{1}{2}, 7\right\}$

Section 6.4

1. $x + 4 = 18$ 3. $x - 4 = 9$
5. $12 = x - 4$ 7. $x + 6 = 12 - x$

9. $x + (x - 4) = 46$
13. $x + (x + 1) = 63$
17. $(x - 2) + x + (x + 2) = 57$
21. $x + (3x + 2) = 75$
25. $2x + (x + 4) = 85$
11. $x + (x + 4) = 18$
15. $x + (x - 2) = 46$
19. $4x = 44$
23. $5(x + 1) + 2x = 68$
27. $x + (x + 4) + (2x - 3) = 46$

29. $25x + 10(14 - x) = 290$ or $\dfrac{x}{4} + \dfrac{14 - x}{10} = 2.90$

31. $60 \cdot x = 40(x + 2)$

Section 6.5

1. True
3. True
5. True
7. False
9. Conditional inequality
11. Conditional inequality
13. Conditional inequality
15. Conditional inequality
17. $x > 3$
19. $x > \dfrac{7}{3}$

21. $x < 6$
23. $x > -3$

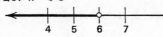
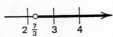

25. $x > \dfrac{-1}{14}$
27. $x > -2$

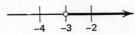

29. $2x > 26$
31. $\dfrac{2x}{3} < 9 - \dfrac{x}{2}$

33. $x + (x + 2) < 87$
35. $8h + 3c > 4600$

Section 6.6

1. $4 + 7$
3. $-(-3)$
5. $2x - 8$
7. $(9 - 4)^2$
9. $(2x + 7) - 3y$
11. $-x^2$
13. $x - 2$
15. (a) $w - 3$
17. (a) $4c - 7$
    (b) $h + 3$
    (b) $\dfrac{b + 7}{4}$

Answers 291

19. $30u + u$ 21. True   23. False
25. False   27. False
29. Conditional inequality   31. Identity
33. Conditional inequality   35. Identity
37. $\{4\}$ 39. $\{-3\}$ 41. $\{6\}$

43. $\{3,4,5,\ldots,10\}$ 45. $\{x \mid x > -4, x \in R\}$

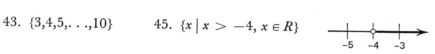

47. 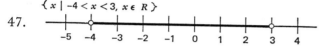   49. $3x + 2 = 9$

51. $x + (x - 1) = -31$ 53. $x - 2 \cdot 5 \leq 120$

Chapter 7

Section 7.1

1. $<<|||$ 3. $|<<||||$ 5. 43   7. 93
9. 4,271   11. 41_5 13. $4,213_5$ 15. $2,043_5$
17. $20,300_5$ 19. 19   21. 32   23. 194
25. 5   27. 265   29. 34_5 31. 143_5
33. $1,042_5$ 35. $11,111_5$

Section 7.2

1. 40_5 3. 101_5 5. 41_5 7. 302_5
9. $2,134_5$ 11. $1,040_5$ 13. $1,010_5$ 15. $1,202_5$
17. $3,334_5$ 19. $2,221_5$ 21. $34,103_5$ 23. $44,002_5$
25. $20,143_5$ 27. $11,104,110_5$

Section 7.3

1. 31   3. 53   5. 272   7. 310
9. 27   11. 3,027   13. 50   15. 514

17.

+	0	1	2	3
0	0	1	2	3
1	1	2	3	10
2	2	3	10	11
3	3	10	11	12

19.

·	0	1	2	3	4	5
0	0	0	0	0	0	0
1	0	1	2	3	4	5
2	0	2	4	10	12	14
3	0	3	10	13	20	23
4	0	4	12	20	24	32
5	0	5	14	23	32	41

21.

+	0	1	2	3	4	5	6	7	8
0	0	1	2	3	4	5	6	7	8
1	1	2	3	4	5	6	7	8	10
2	2	3	4	5	6	7	8	10	11
3	3	4	5	6	7	8	10	11	12
4	4	5	6	7	8	10	11	12	13
5	5	6	7	8	10	11	12	13	14
6	6	7	8	10	11	12	13	14	15
7	7	8	10	11	12	13	14	15	16
8	8	10	11	12	13	14	15	16	17

23.

·	0	1	2	3	4	5	6	7	8	9	t	e
0	0	0	0	0	0	0	0	0	0	0	0	0
1	0	1	2	3	4	5	6	7	8	9	t	e
2	0	2	4	6	8	t	10	12	14	16	18	1t
3	0	3	6	9	10	13	16	19	20	23	26	29
4	0	4	8	10	14	18	20	24	28	30	34	38
5	0	5	t	13	18	21	26	2e	34	39	42	47
6	0	6	10	16	20	26	30	36	40	46	50	56
7	0	7	12	19	24	2e	36	41	48	53	5t	65
8	0	8	14	20	28	34	40	48	54	60	68	74
9	0	9	16	23	30	39	46	53	60	69	76	83
t	0	t	18	26	34	42	50	5t	68	76	84	92
e	0	e	1t	29	38	47	56	65	74	83	92	t1

27. 102_8 29. 1012_4 31. $1,110_3$ 33. $2,013_9$
35. $881e_{12}$ 37. $1,575_8$ 39. $23,355_6$

Section 7.4

1. 6 3. 12 5. 27 7. 109
9.

+	0	1
0	0	1
1	1	10

11. 110_2 13. 110_2
15. $110,000_2$ 17. $10,001,111_2$ 19. $1,101_2$ 21. $1,000,001_2$
23. $10,001,001_2$ 25. $111,100,011_2$ 27. All end with a 1

Section 7.5

1. << < | | | | | |
   <<
3. 36
5. 321_4
7. $4,004_7$
9. $101,001_2$
11. $t0,003_{12}$
13. 22
15. 54
17. 11
19. 80
21. 33
23. 1,572
25. 1,994
27. 14,409
29. 223_4
31. ee_{12}
33. $1,343_5$
35. 436_7
37. 337_8
39. $221,221_3$
41. 62_7
43. 221_8
45. $10,012_3$
47. $2,122_4$
49. $1,643_9$
51. 434_{12}
53. 210_3
55. $3,012_4$

59.

·	0	1	2
0	0	0	0
1	0	1	2
2	0	2	11

Index

absolute value, 46*ff*, 53, 54, 62, 63, 82
addition
  addends of, 13
  associative property of, 15
  commutative property of, 14
  identity element for, 16
  mixed operations involving, 26, 58, 64, 99, 115
  of integers, 48*ff*
  of rationals, 109*ff*, 123
  of whole numbers, 12, 13
additive inverse, 50, 52, 78, 98, 123
algebraic expressions, 159, 205*ff*, 209
algorithm
  addition, 17*ff*, 51*ff*, 182
  changing base, 241*ff*, 252
  definition of, 19
  multiplication, 30, 180, 248
approximations
  decimal, 189*ff*
  symbol for, 190
associative property
  of addition, 15, 54, 71, 115, 245
  of multiplication, 27, 61, 78, 249
  of union operation, 15
average, 191

Babylonian numeral system, 238
base
  of exponent, 130
  of numeral system, 237
binary
  numeral system, 257
  operation, 12

calculate, 12
cardinality, 5
closed, 26
closure
  definition of, 26
  with respect to addition, 26, 71
  with respect to subtraction, 26
commutative property
  of addition, 14, 54, 71, 115, 245
  of multiplication, 27, 61, 78, 247
  of union operation, 14
completely factored, 107
composite number, 107
compound fractions, 120*ff*
  main fraction bar of, 97
consecutive, 208*ff*
constants, 159, 207
correspondence, 239
cube, 130
  perfect, 142
  root, 141

data processing, 262
decimals
  approximations for irrational numbers, 189*ff*
  decimal point, 173
  nonrepeating, nonterminating, 190
  nonterminating, 176
  numeral system, 237
  operations with, 179*ff*
  repeating, nonterminating, 176, 196
  terminating, 176
denominator, 81, 97, 110, 111
  irrational, 152
  least common (LCD), 108
  rationalizing the, 152*ff*
dense, 198
difference, 22, 52, 98, 113
digit, 8
digital computer, 261
directed distance, 46

distance, 46
distributive property, 66*ff*, 123
  over addition, 66
  over subtraction, 67
dividend, 96
division
  exact, 105
  long, 104
  mixed operations involving, 99
  of rational numbers, 96*ff*
  operation of, 12
  rule, 102*ff*
divisor, 96
domain, 207
double negative, 50
duodecimal numeral system, 256

Egyptian hieroglyphic numeral system, 5, 237
equalities, 33
equation, 215*ff*
  application of, 220
  conditional, 217
  solving an, 218
even, 130
exactly divisible, 105
expanded notation, 18, 132, 240
exponential notation, 129*ff*
  laws of, 134, 135, 138
exponents, 130
  negative, 167
  nonpositive, 167*ff*
  operations with, 137*ff*
  zero, 170
expressions
  algebraic, 159, 205, 209
  mathematical, 206
  numerical, 209

factor, 68, 88, 176
  common, 68, 88, 91*ff*, 149
  greatest common (GCF), 68, 88
factoring, 68
  complete, 107
field, 198
  ordered, 198
  properties of, 197
fractions, 81
  common, 173, 178

fractions (*continued*)
  compound, 120*ff*
  decimal, 173*ff*
  denominator of, 81
  equivalent, 88
  fundamental property of, 84*ff*, 88, 149
  numerator of, 81
  reduced, 88
  rules of signs of, 86
  simple, 120
  unit, 81

glossary, 269
graph, 17
greatest common factor (GCF), 87, 88, 107, 124

hexadecimal, 261
Hindu-Arabic numeral system, 8, 132, 237

identity, 217
identity element
  for addition, 16, 78, 123
  for multiplication, 31, 64, 123
  for union operation, 15
indeterminate, 103
index (for root), 141
inequalities, 33*ff*, 196, 226*ff*
  absolute, 226
  conditional, 226
  simple, 36
  single, 36
integers, 44*ff*
  addition of, 48*ff*
  consecutive, 208
  mixed operations with, 58
  multiplication of, 61*ff*
  negative, 45
  nonnegative, 45, 143
  nonpositive, 45
  positive, 44, 45
  subsets of, 45, 76
  subtraction of, 56*ff*
inverse
  additive, 50, 78
  multiplicative, 84
inverse operations, 58, 99
irrational numbers, 129*ff*, 142, 196

irrational numbers (*continued*)
  decimal approximations for, 189*ff*
  definition of, 189, 190
  evaluation of, 157*ff*

least common denominator (LCD), 108, 118, 124
least common multiple (LCM), 107, 108, 111, 124

mathematical expressions, 206
mathematical sentences, 205
mathematical statements, 205
Mayan numeral system, 239
minuend, 22
minus sign, 22
mixed operation
  definition of, 25
  rule for, 26, 58, 64, 99, 115
multiple, 106
  least common (LCM), 106*ff*
  common, 106
multiplication
  algorithm for, 29
  defined, 27
  identity element for, 31, 64
  mixed operation involving, 32, 64, 99
  of integers, 61*ff*
  of rationals, 75*ff*, 123
  of whole numbers, 27
  operation of, 12
multiplicative inverse, 77, 84, 98, 123

negative, 44
  double, 50
  non-, 45, 143
number line, 17
numbers
  composite, 107
  counting, 10
  double negative of, 50
  even, 130
  irrational, 129*ff*, 142, 189, 196
  mixed, 104
  natural, 10
  odd, 130
  ordinal, 130
  power of, 130

numbers (*continued*)
  prime, 107
  rational, 76*ff*
  real, 195*ff*
  whole, 10
numeral systems
  Babylonian, 238
  base five, 240*ff*
  base three, 252*ff*
  binary, 257
  decimal, 237, 256
  duodecimal, 256
  Egyptian hieroglyphics, 5, 237
  hexadecimal, 261
  Hindu-Arabic, 8, 132, 237
  Mayan, 239
  octal, 261
  positional, 8*ff*
  Roman, 6, 237
  simple grouping, 6
numerals, 5
  face value, 8
  place value, 8
numerator, 81, 97, 110, 111
numerical expressions, 209

octal, 261
odd, 130
open sentence, 216
  roots of, 217
  solutions of, 217
operations
  addition, 12, 246
  binary, 12
  division, 12, 96
  inverse, 58, 99
  mixed, 25, 26, 58, 64, 99, 115
  multiplication, 12, 248
  subtraction, 12
  unary, 140
  union, 12
  with base five, 244*ff*
  with base three, 252*ff*
  with decimals, 179*ff*
  with radicals, 148*ff*
opposite, 44
order
  defined, 34, 35
  (-)ed field, 198

order (*continued*)
  of integers, 47
  of rational numbers, 117*ff*
  of real numbers, 198
  of whole numbers, 33
ordered field, 198
ordinal number, 130

perfect
  cube, 142
  power, 142
  square, 142
pi (π), 190
plus, 44
polynomials, 158
positional, 9
power, 130
  negative, 167
  perfect, 142
  zero, 170
prime
  number, 107
  relatively, 107
product
  of integers, 62
  of rational numbers, 82, 102
  of whole numbers, 27
Pythagorean theorem, 158

quotient, 96
  partial, 105
  rule for, 98

radical, 139*ff*
  like, 148
  operations with, 148*ff*
  sign, 140
  unlike, 148
radicand, 140
rational numbers, 76*ff*
  addition of, 109*ff*
  division of, 96*ff*
  multiplication of, 75*ff*, 81
  order of, 117*ff*
  subsets of, 76
  subtraction of, 109*ff*
real number system, 195*ff*
  real number, 196
  subsets of, 197

reciprocal, 77, 84, 168
relatively prime, 107
relations, 36
remainders, 104
replacement set, 207
Roman numeral system, 6, 237
root
  cube, 141
  index of, 141
  principle square, 140
  square, 140
  table, 267
rules
  for addition, 53
  for division, 98
  for multiplication, 62, 63

satisfies, 217
scientific notation, 184*ff*
sentences
  grammatical, 231
  mathematical, 205
  open, 216
  roots of, 217
  satisfies a, 217
  solutions of, 217
set builder notation, 227
sets
  belongs to, 4
  disjoint, 13, 196
  elements of, 4
  empty, 10
  finite, 10
  infinite, 10
  members of, 4
  null, 10
  replacement, 207
  solution, 217
  truth, 217
  union of, 12
simple grouping systems, 6
statements, mathematical, 205
subscript, 239
subset, 45, 76, 123, 196
subtraction
  algorithm for, 24
  definition of, 22
  mixed operations involving, 26, 58, 99
  of integers, 56*ff*

subtraction (*continued*)
  of rationals, 109*ff*
  of whole numbers, 21*ff*
  operation of, 12
  rule for repeated, 25
subtrahend, 22
sum, 13

unary operation, 140
undefined, 85
union
  associative property of, 15
  commutative property of, 14
  identity element of, 15
  of sets, 12

value, 207
  absolute, 46
  face, 8
  place, 8
variable, 158, 207

whole numbers, 10*ff*, 76
  subsets of, 45

zero
  as exponent, 170
  as identity element of addition, 16
  as placeholder, 9
  factor property, 31, 71